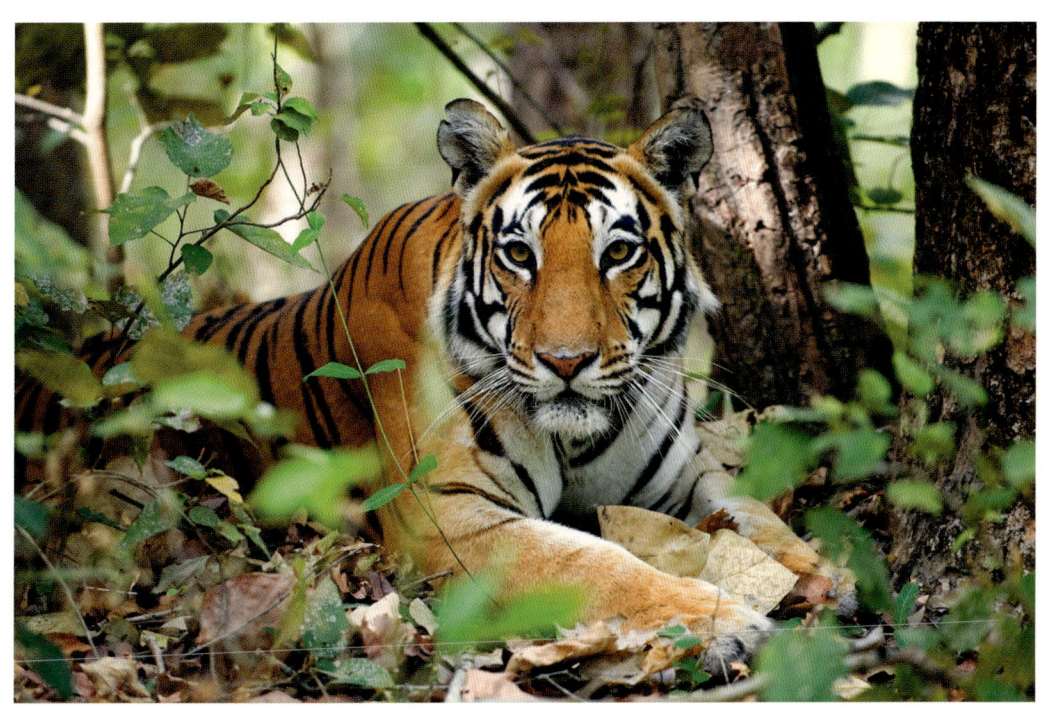

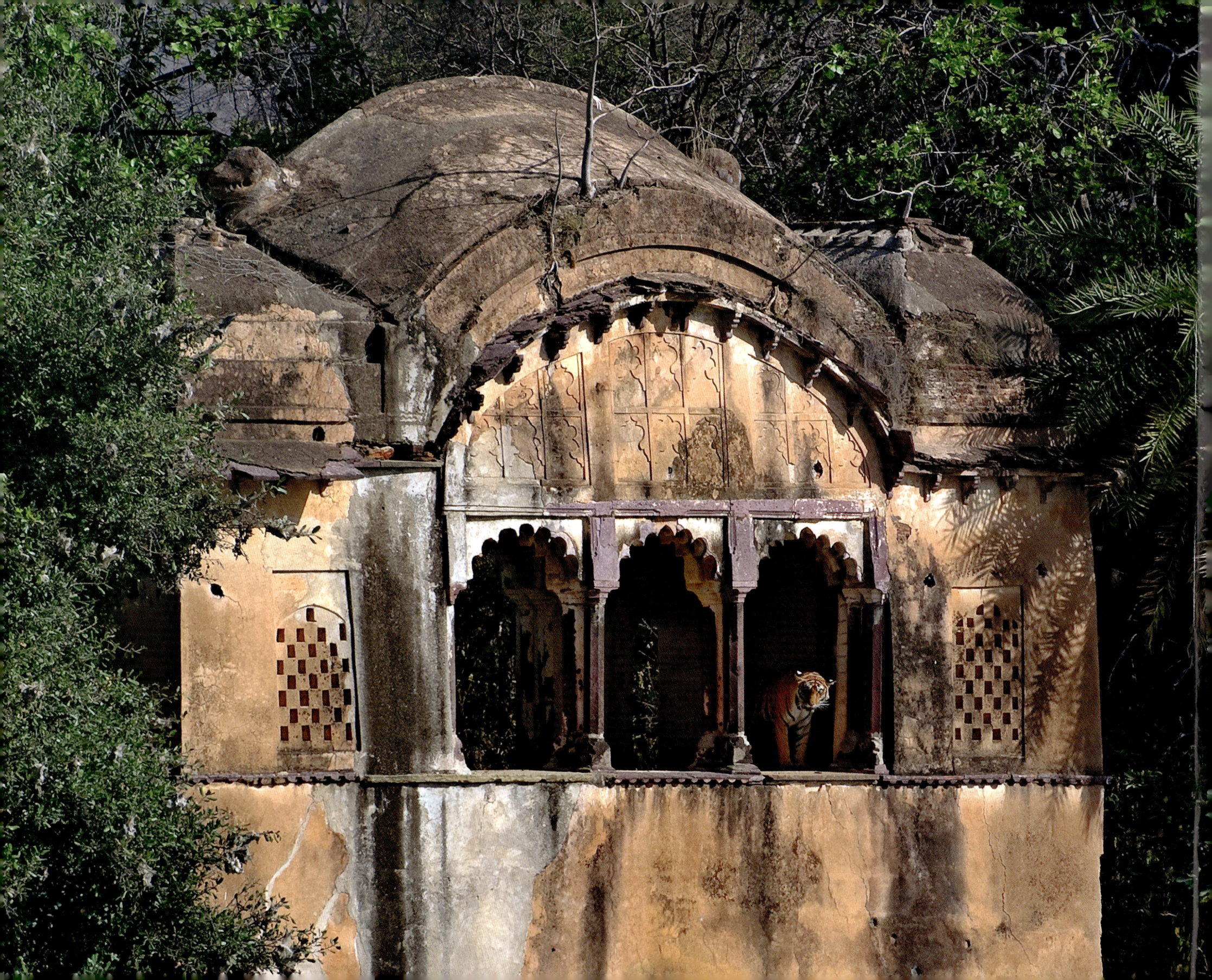

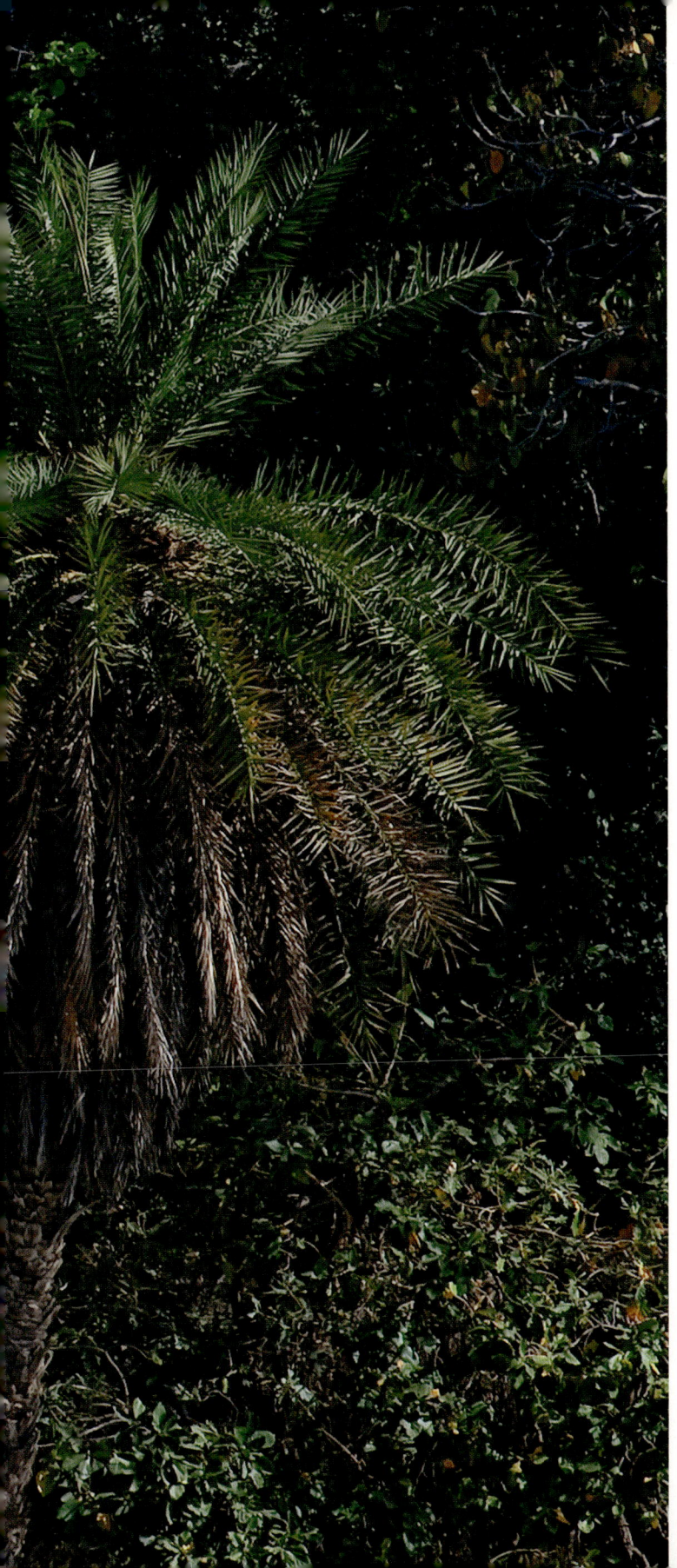

AXEL GOMILLE

INDIA

LAND OF TIGERS AND TEMPLES

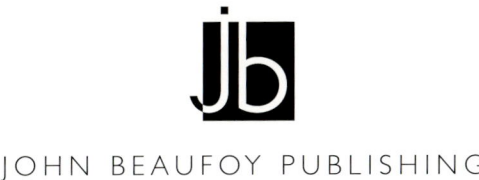

JOHN BEAUFOY PUBLISHING

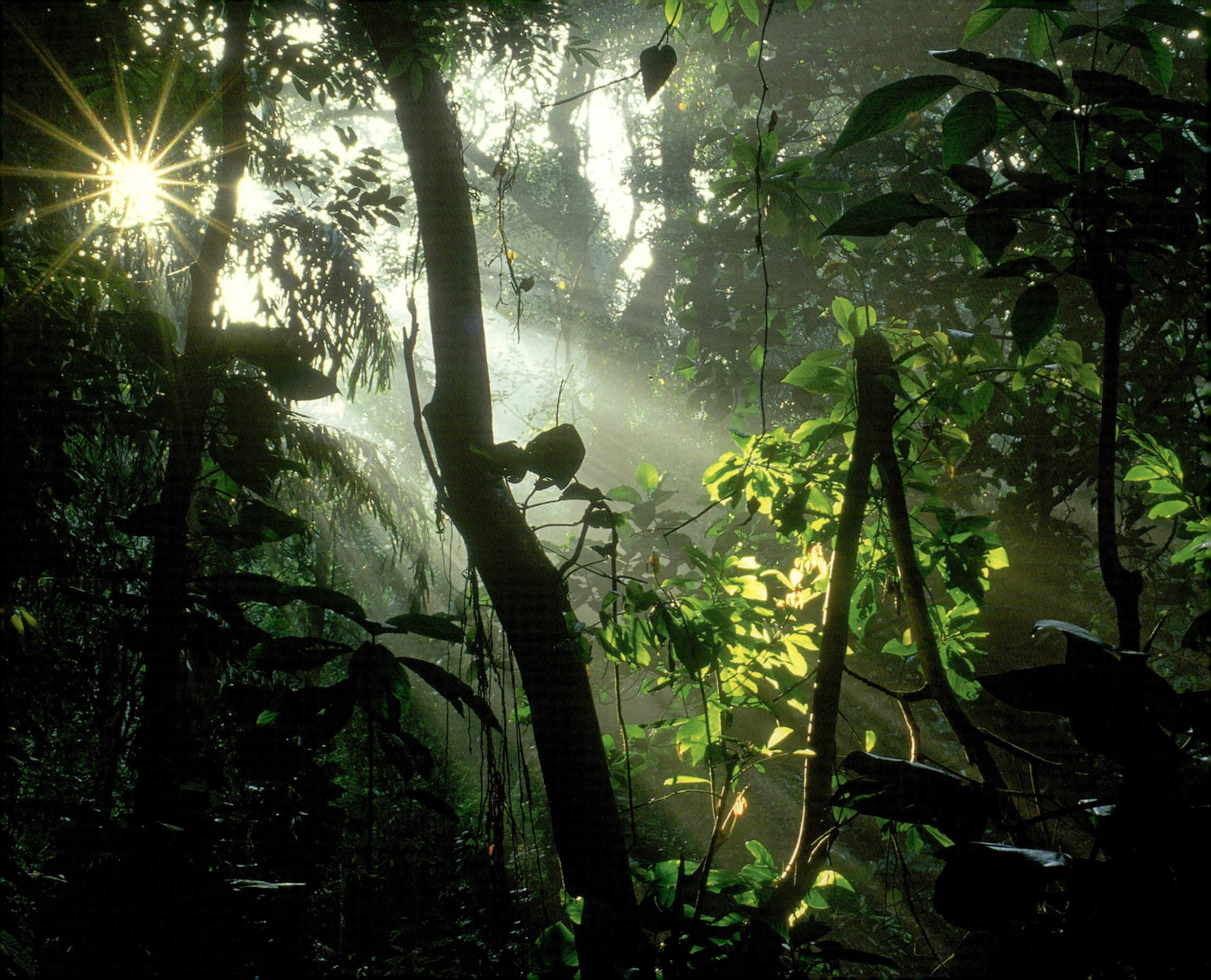

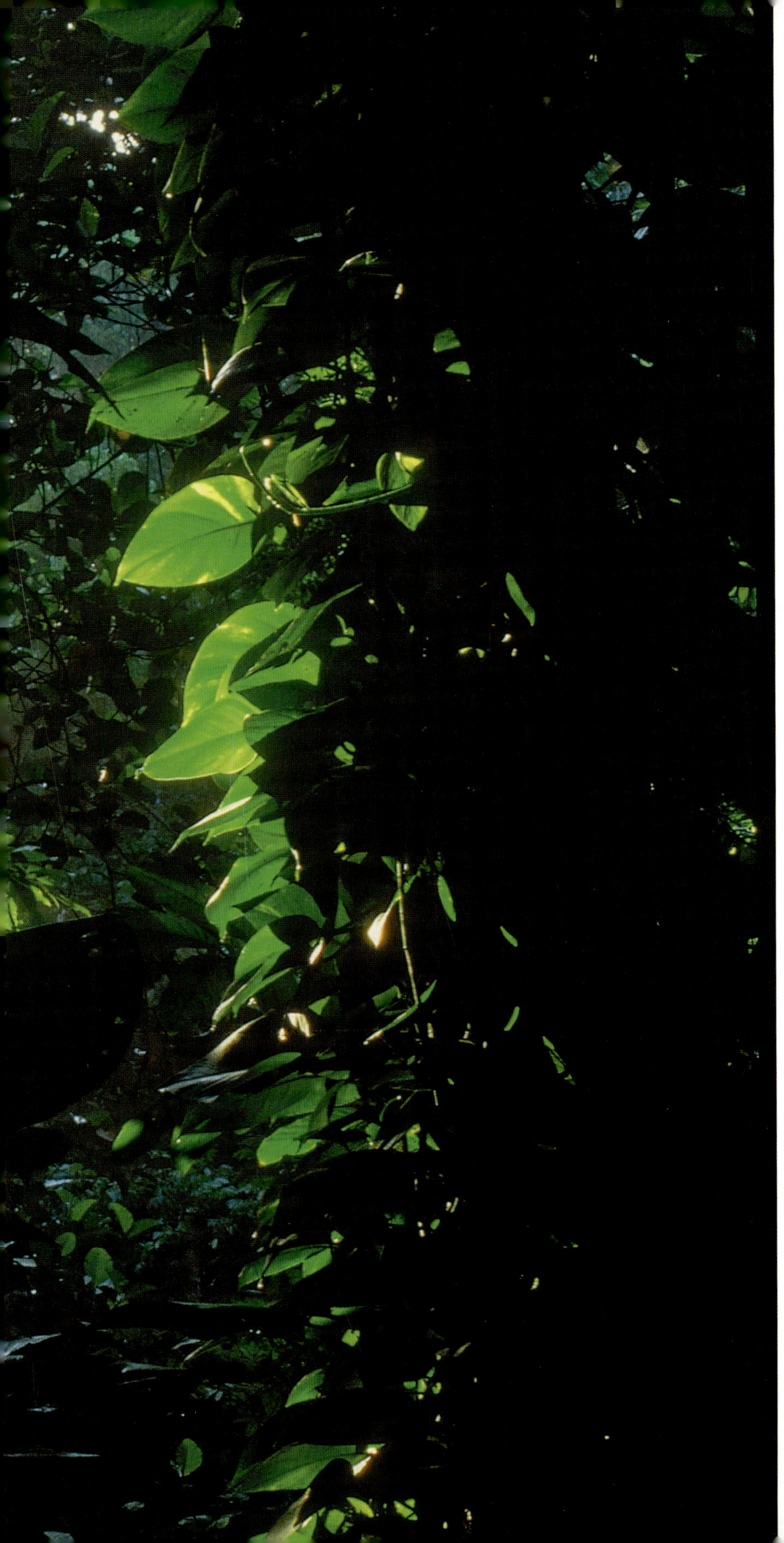

CONTENTS

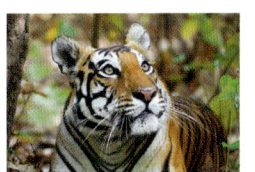
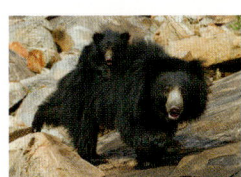
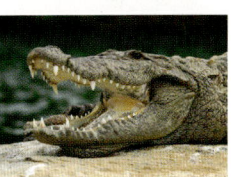
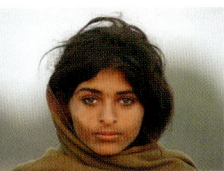
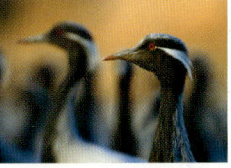
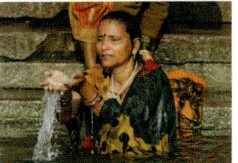
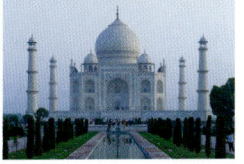

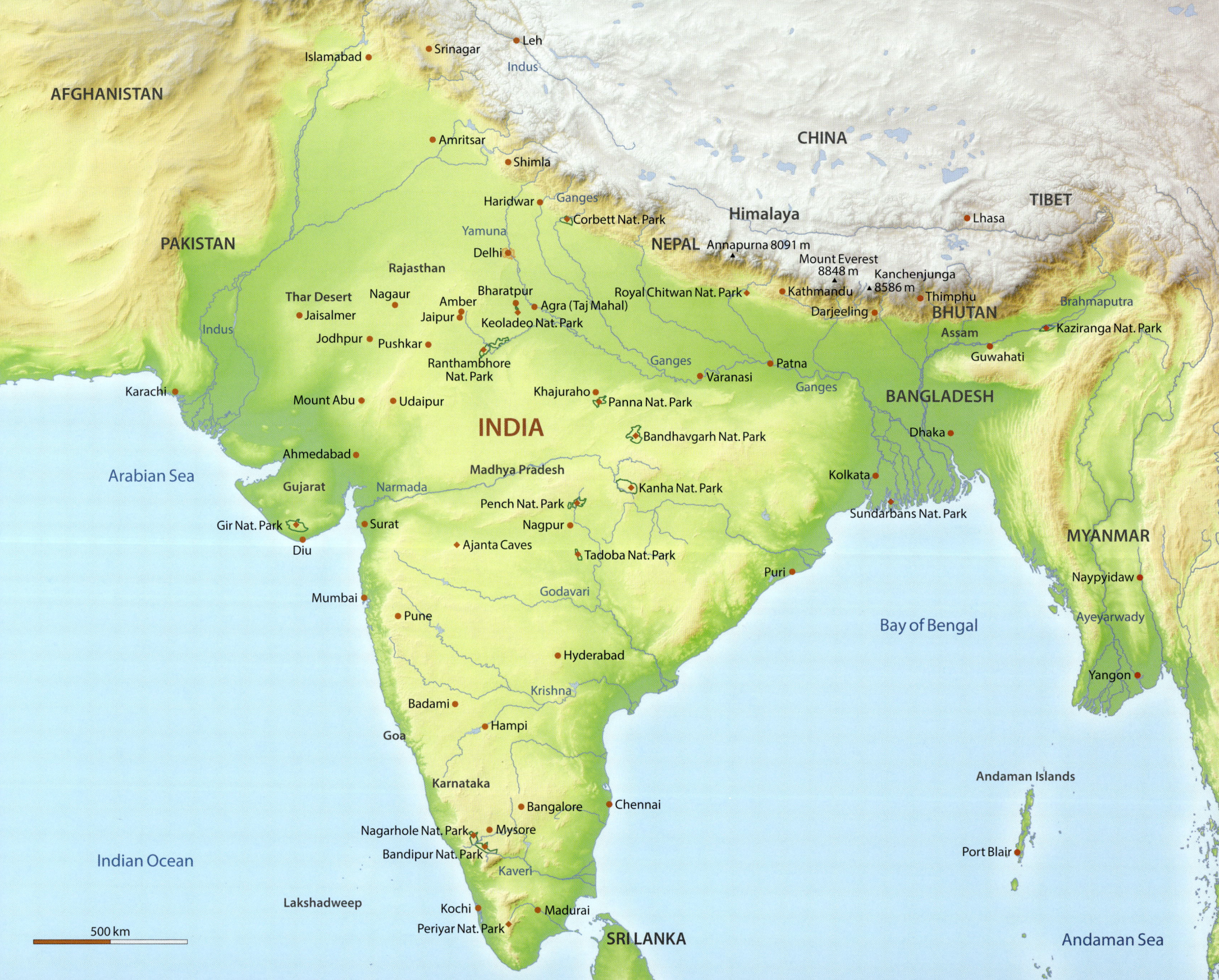

Introduction

How many Indian megacities can you name? Delhi, Kolkata and Mumbai probably come to mind, but beyond those three? The five largest Indian metropolises have a combined population figure of roughly 65 million – about the same as the whole of the United Kingdom. And India has a further 41 cities with a population of more than a million – Pune and Surat, for example. Both are larger than either Berlin or Madrid, but most of us have never even heard of them. And then there are many other large and small towns, and an almost endless list of villages. The current population of India is estimated at about 1.2 billion people and rising. In just a few years India could overtake China to become the most populous country on earth.

Amongst all these people, is there even any space left for wildlife? The surprising answer is: most certainly. Today in India you will see parakeets in gardens everywhere, kingfishers on the power supply lines, monkeys on roofs and antelopes out in the fields. Wild animals are more or less omnipresent. This kind of diversity in such a heavily populated country is truly special.

The Indian subcontinent has several biodiversity hotspots, where the abundance of plants and animals is even higher than average. Of the world's 37 known species of wild cats, 14 are found in India – more than in any other country. In contrast, there are only ten different species of wild cats in the whole of Africa. This is just one example of the importance of India as a habitat for wild fauna and flora. There are about 350 species of mammals, 540 species of reptiles and 1,200 species of birds breeding in India.

One reason for this diversity lies in the geography of the Indian subcontinent. The landmass is enormous. The distance from the beaches of Tamil Nadu, at the southernmost tip of India, to the snow-capped mountains of the Himalayas in the north is around 3,200 kilometres (2,000 miles). And it is almost as far from the Thar Desert in the west, near the Pakistani border, to the rainforests in the northeast on the border with Myanmar. The variety of habitats encompassed within this vast space has led to an extremely diverse flora and fauna.

But how does this diversity survive today, when in so many other places on earth it has been lost? The answer lies in the people and how they treat their fellow creatures. Hinduism, the dominant religion of India, preaches respect for all forms of life. In the countryside especially, where people and wild animals come into contact most often and where religion is part of everyday life, the wildlife benefits from these religious principles. Many animals, such as monkeys, elephants and cobras, are even considered to be holy.

These traditional values still have a place. India is a country of contrasts: beautiful and ugly, rich and poor, modern and medieval. I was fascinated – and particularly wanted to spend more time in the jungle, so I worked as a naturalist in the tiger reserves of Kanha and Bandhavgarh. Over a period of more than 20 years I travelled as often as I could through various regions of the Indian subcontinent. This book is one of the results. It is not a comprehensive account, but rather a personal view of India, one of the most interesting, most beautiful and most diverse countries on earth – the land of tigers and temples.

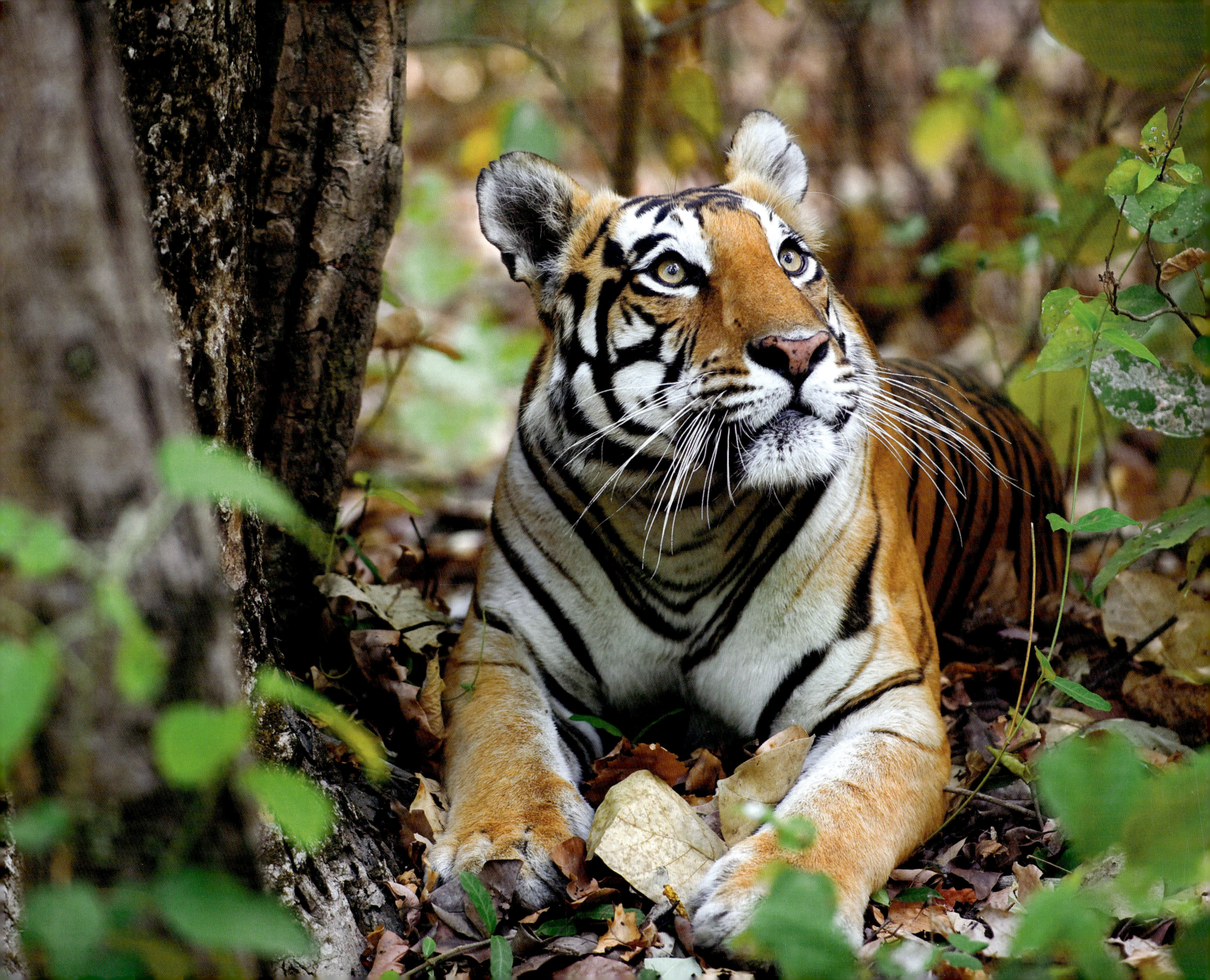

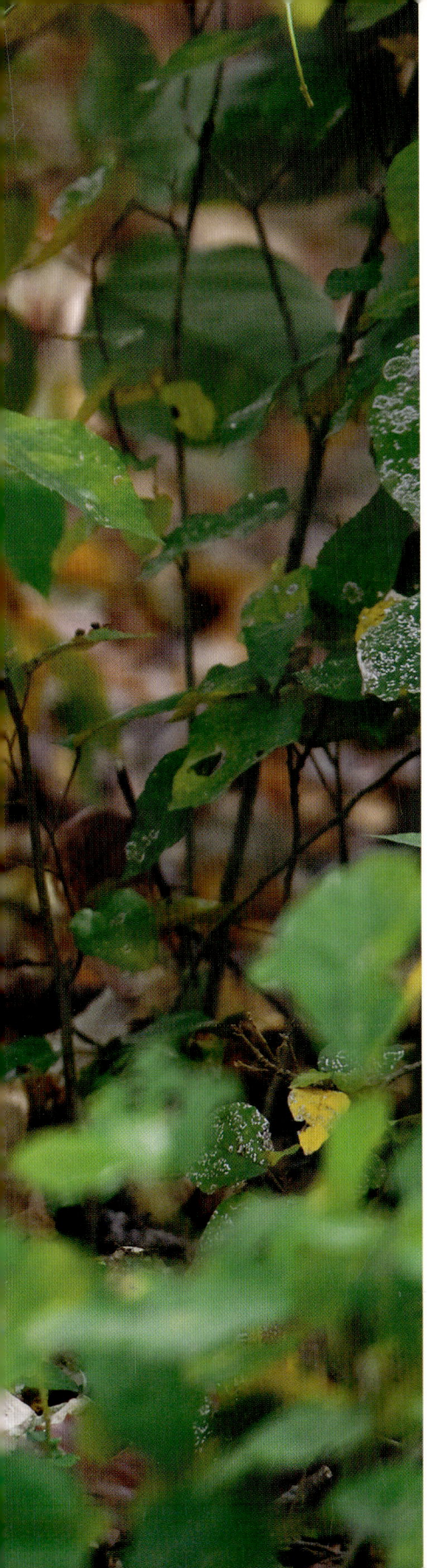

In the Realm of the Tiger

India

To look a wild tiger in the eye is one of the highlights of travelling through India. Where this big cat roams the jungle, nature is still largely intact.

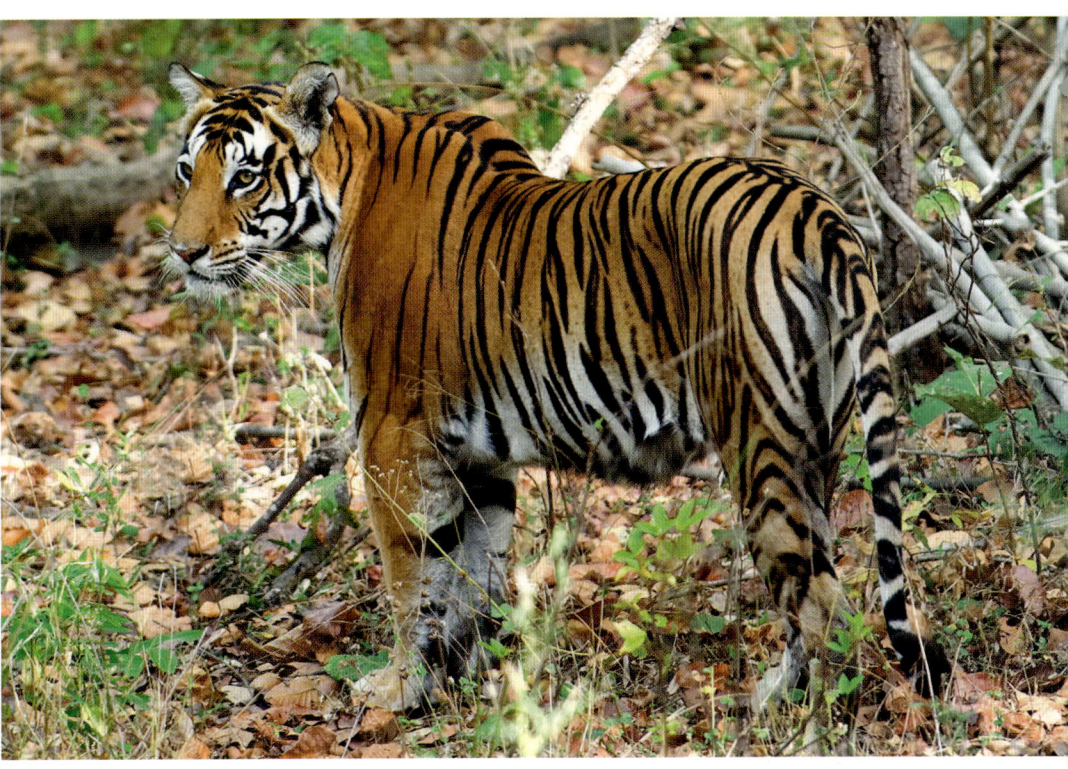

Spirit of the Indian Jungle

Up in the trees, excited langur monkeys scream and leap from branch to branch. On the ground, stately sambar deer utter their barking alarm calls. Their raised tails show that they are alert, eyes and ears fixed on something out there in the jungle. The animals become agitated, the tension rises. Finally, the leaves at the edge of the forest stir a little and a striped body slips through the tall grass and pauses in a clearing. Powerful, majestic and beautiful – this is indeed the lord of the jungle.

No animal embodies the spirit of the Indian jungle like the tiger. When a tiger appears, hardly anyone is immune to the magic of the moment. Whether local villager or international traveller, irrespective of origin, age or cultural background – the big cat seems to enchant everybody. However, you need a fair bit of luck and persistence to encounter a tiger in the wild, as they are usually shy and have now become very rare.

In the past, seeing a tiger would have been easier, as the natural range of the species covered much of Asia. This adaptable big cat was equally at home in the reed beds of the Caspian Sea and in eastern Siberia with its ice-cold winters, in the mountains of eastern Turkey and in the Indian scrub jungle or the rainforests of Indonesia. Hunting, poaching and the destruction of their habitat led to a major decline in tiger range and population. Out of eight tiger subspecies, three have been driven to extinction within the last century, and those that remain have a highly fragmented distribution. In all of Asia, wild tiger populations have decreased by a shocking 96 per cent – to around 4,000 individuals. About half of these live in India, more than in any other country within their range. That is why India is of such vital importance for the survival of the species.

However, the results of a tiger census in 2005 were even more alarming. In Sariska National Park the last tigers had been poached under the very eyes of the rangers and the park authorities – a new low point in the efforts to conserve India's national animal. Soon afterwards, Panna National Park lost the last of its tigers as well. In spite of the veneration for life that is so pervasive in India for religious reasons, some people just cannot resist the economic incentive to poach. This downward trend had to be stopped. The reorganization of the national authority responsible for tiger conservation has given new momentum to these efforts. The creation of new reserves and the strict surveillance of protected areas were important steps in this direction. Today, there are 43 tiger reserves in India, protecting 65,000 square kilometres (25,000 square miles) of important tiger habitat – an area about the size of Sri Lanka.

Considering the dramatic decline of the tiger, are the efforts to save it worthwhile? To answer this question, it is helpful to look back at recent history. When Project Tiger, the first major effort to save the species, was launched in 1973, there were about 1,800 tigers in India. But thanks to strict conservation measures over the next 20 years, the population more than doubled, according to official figures. The tiger population thus has the potential to recover quickly. The cats need space, water and prey animals – and with respect to the latter they are not choosy. The number of wild tigers has plummeted, but there are still remnant populations in many different areas. If poaching can be prevented and habitat restored, the number of these majestic big cats could rise again – and such an outcome would surely be worth the effort!

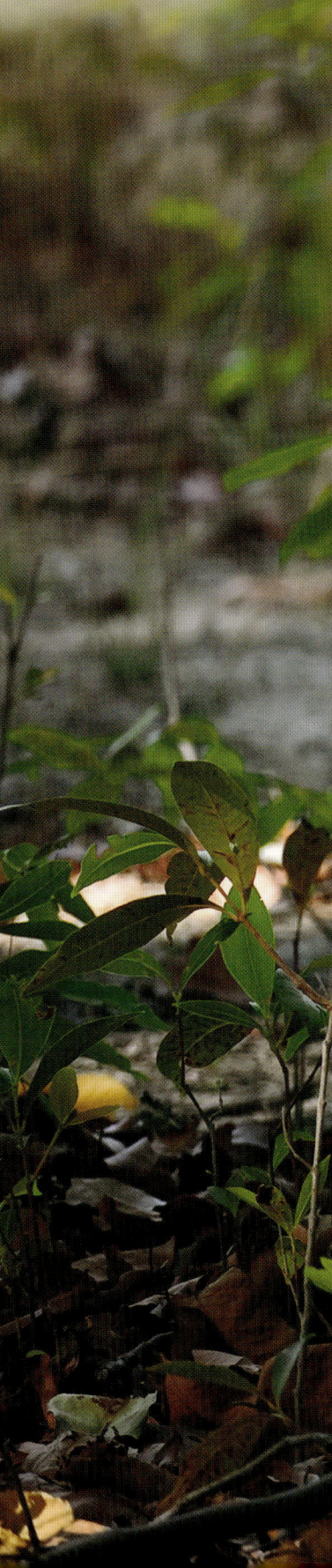

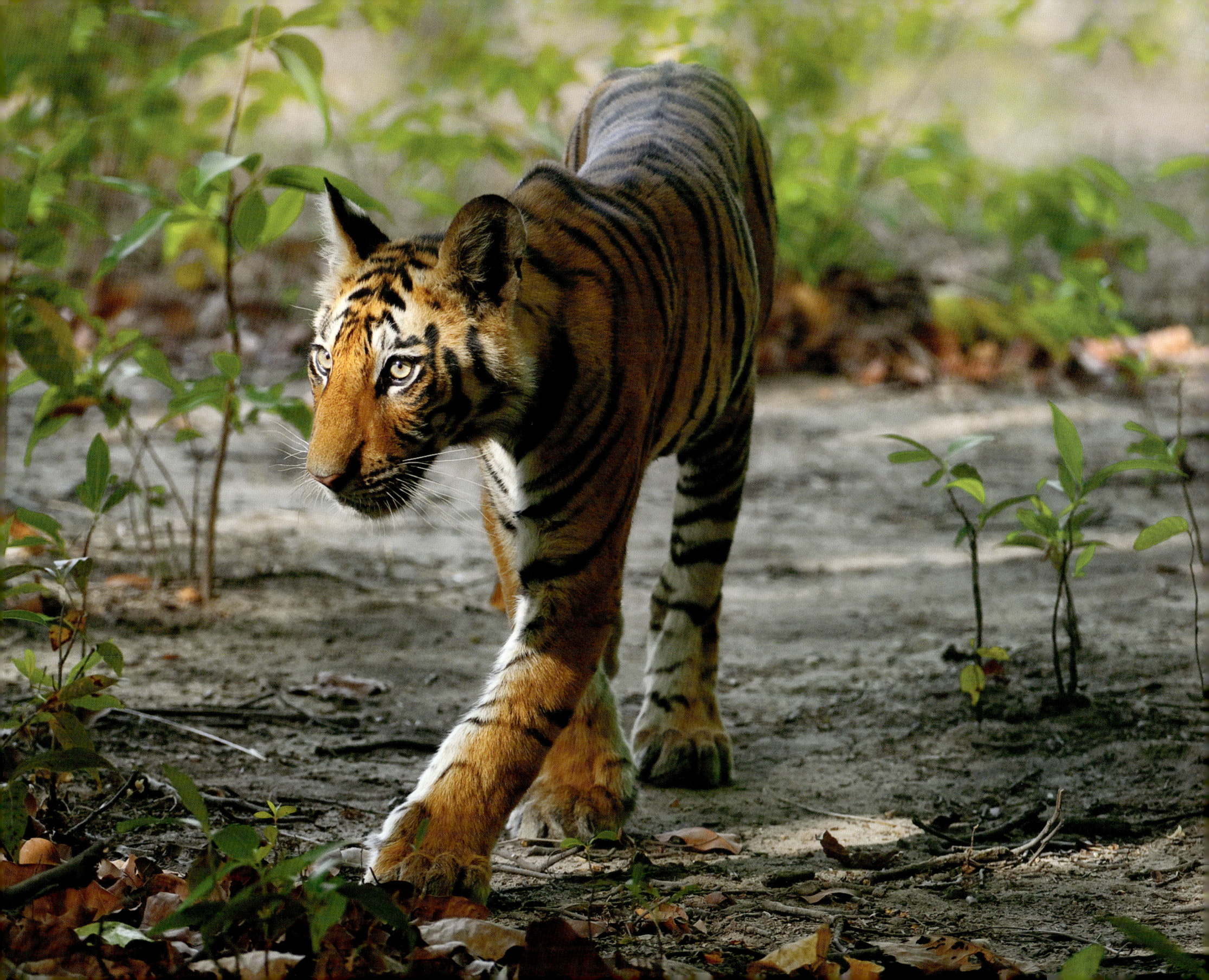

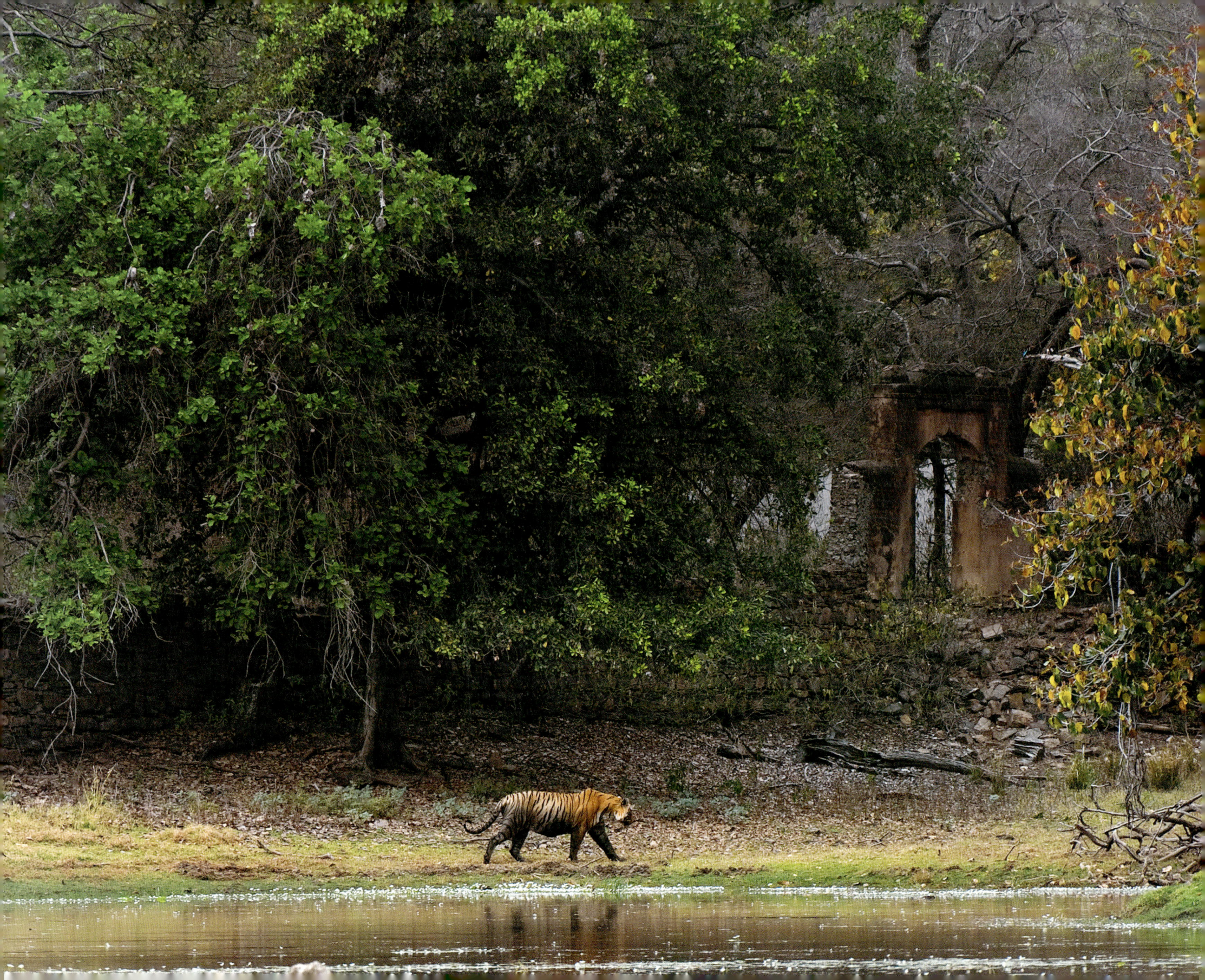

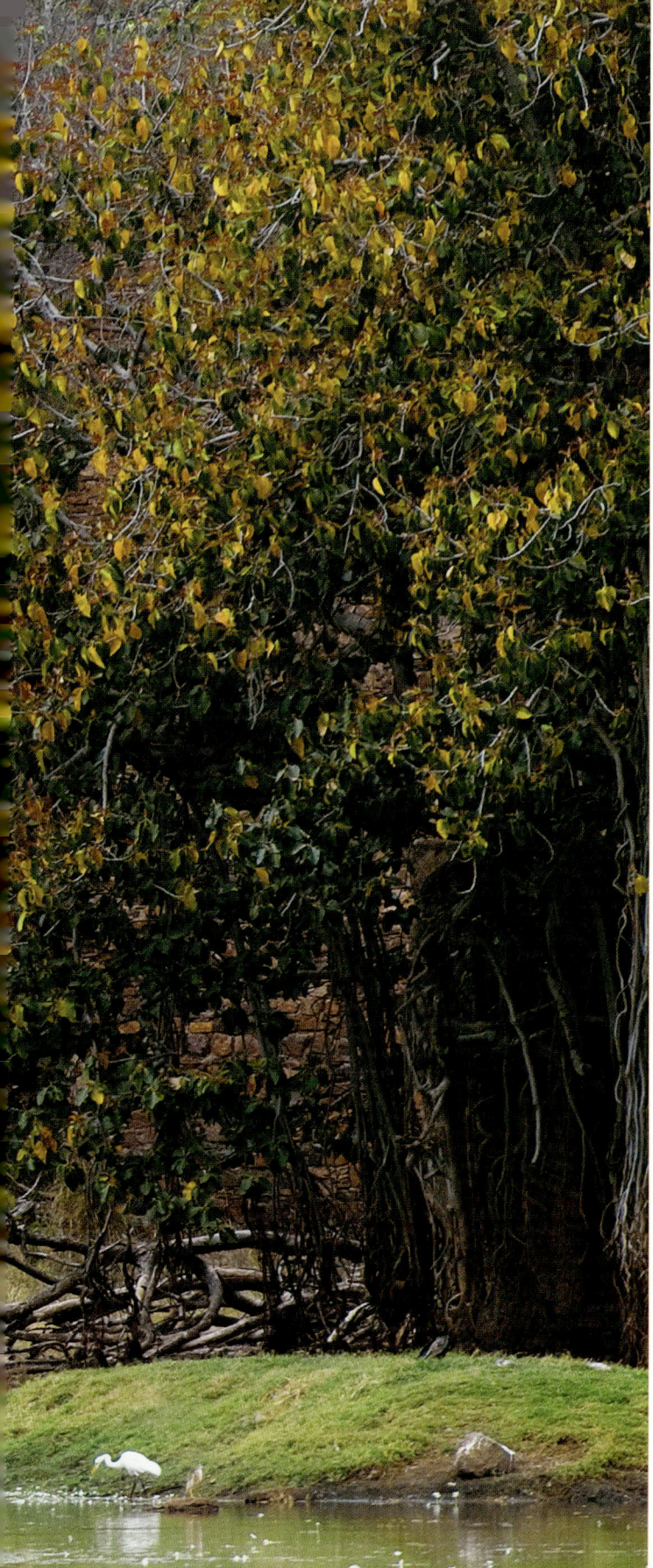

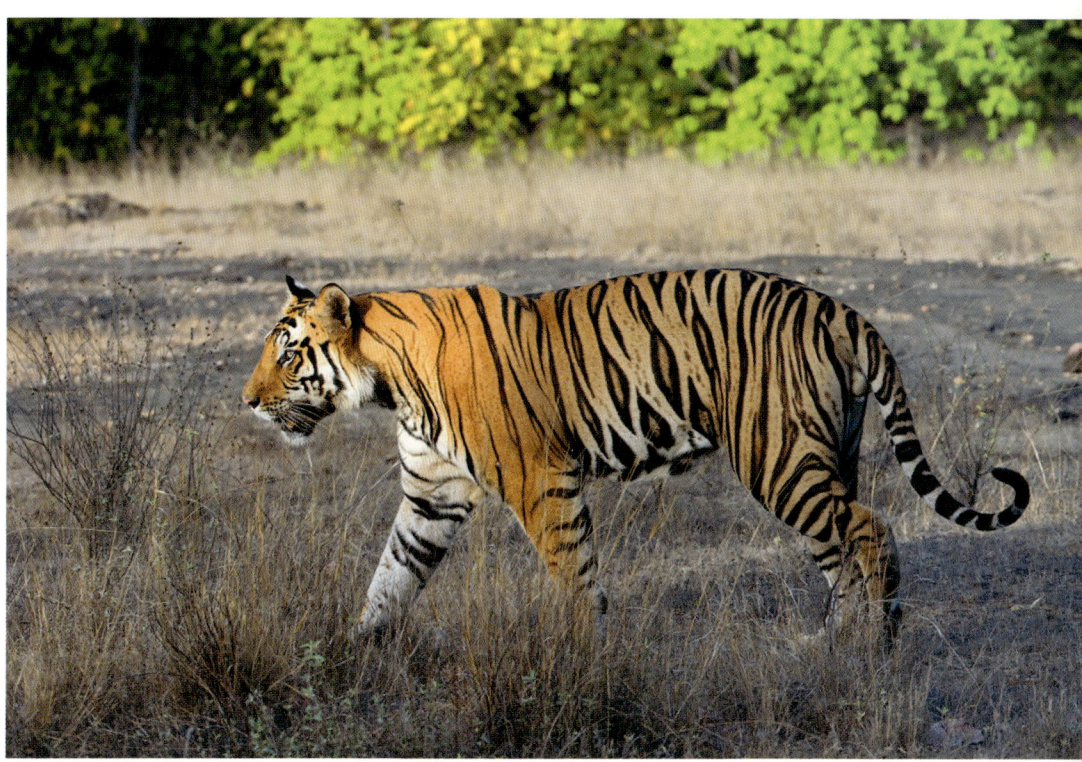

A male tiger roams Ranthambore National Park in Rajasthan. Ruins of temples and palaces, reclaimed by the jungle, have become a part of the tiger's territory. This reserve is famous for offering a good chance of encountering the big cats out in the open.

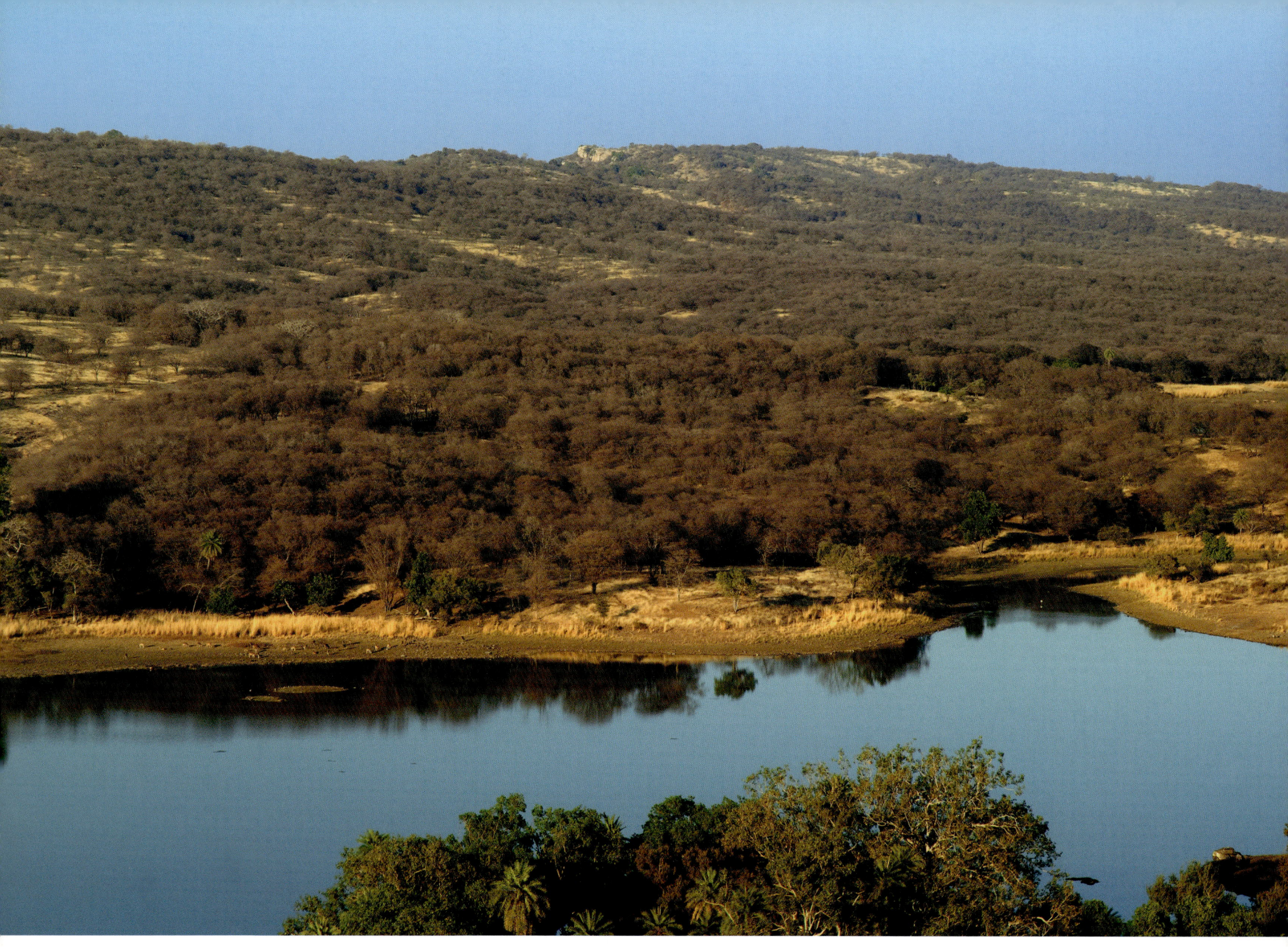

In the heart of Ranthambore National Park lie a number of lakes. These areas consistently support a high density of prey species, and are often frequented by the tigers.

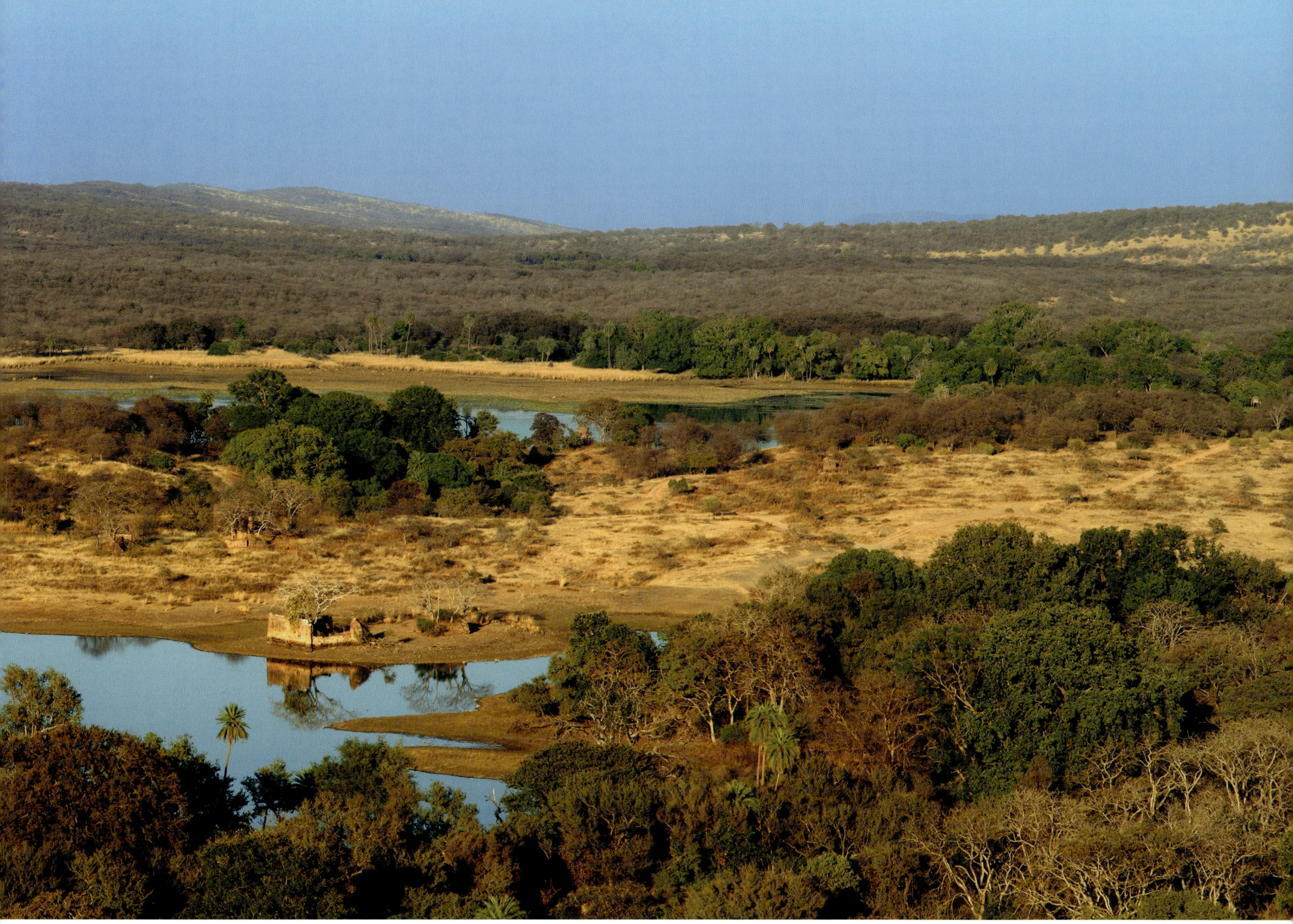

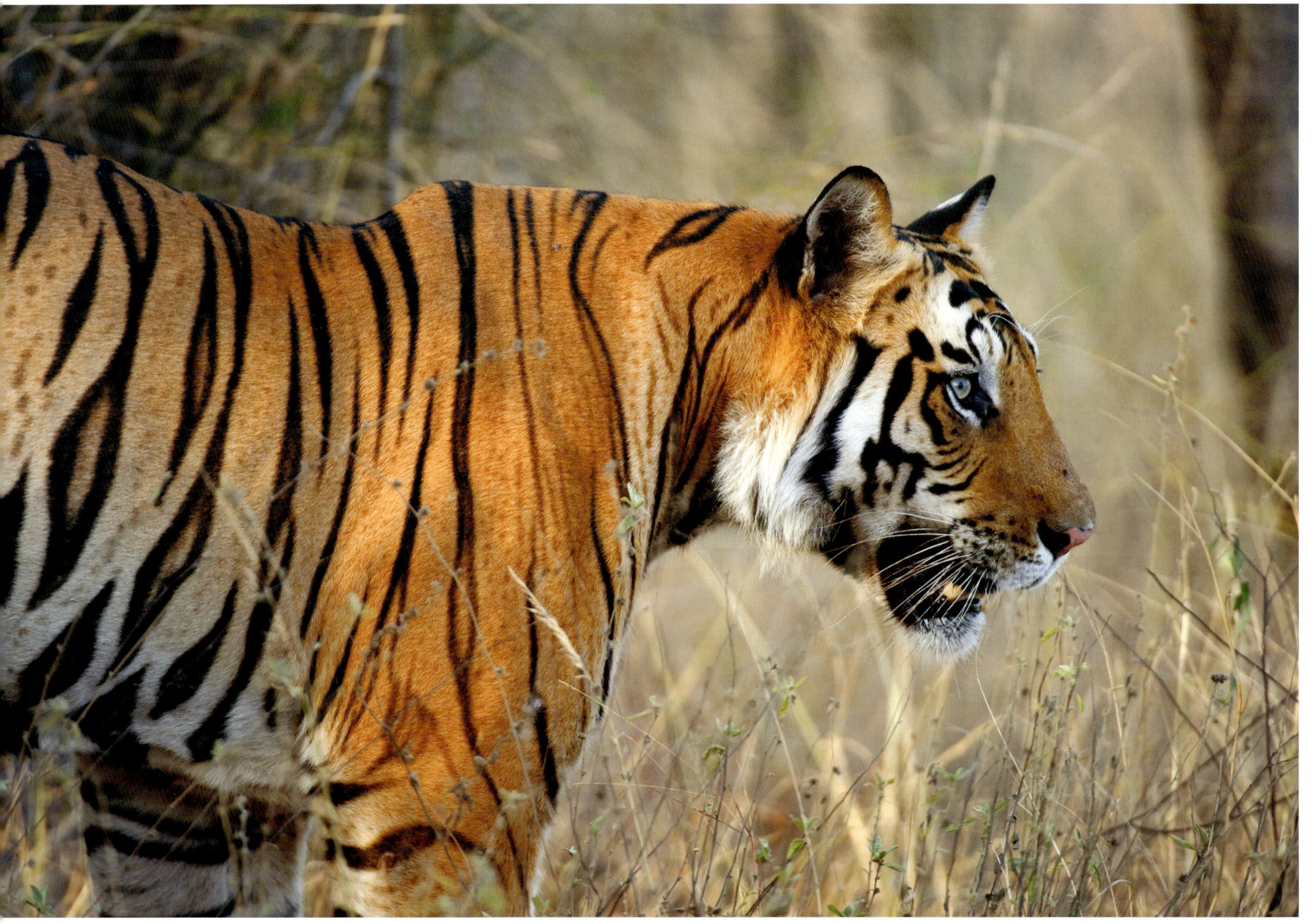

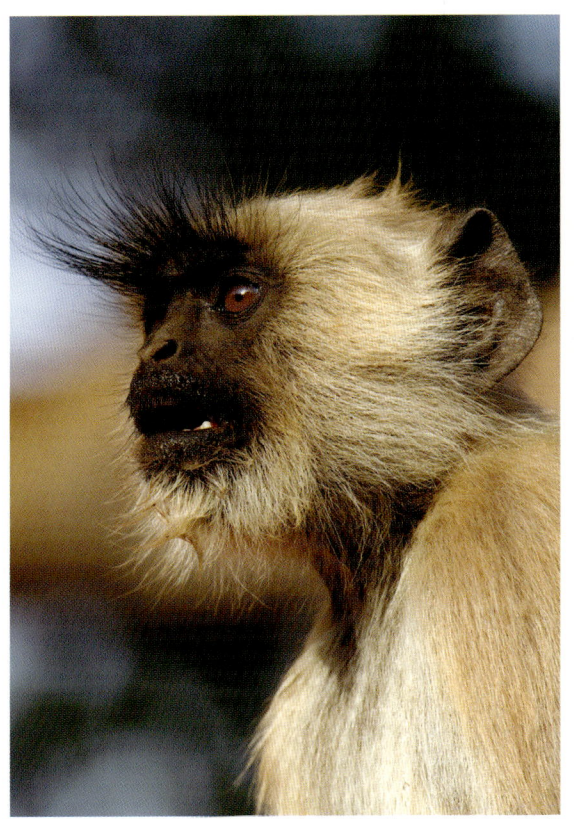
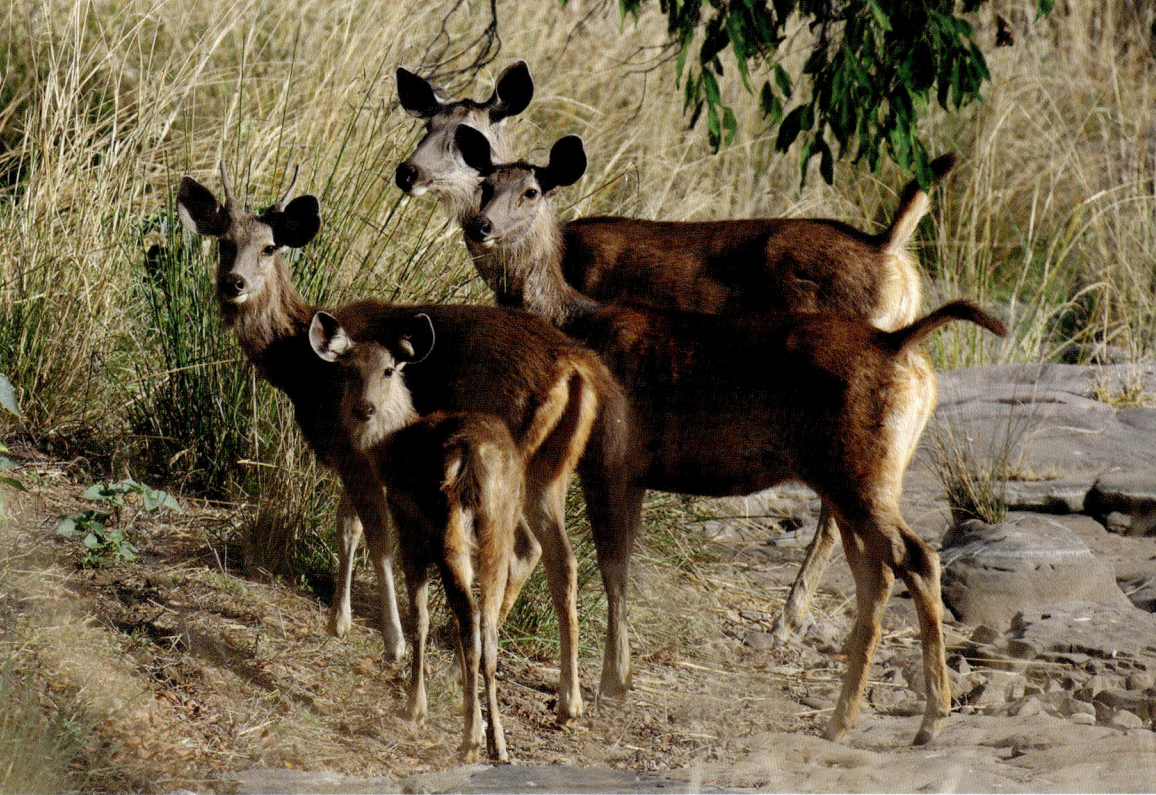

A tiger on the prowl spells danger for many animals. Several jungle dwellers therefore give characteristic alarm calls when a tiger is seen. From their vantage points high up in the canopy, Hanuman langurs are often the first to spot the predator. When the langurs start calling, the sambar deer are immediately alert, with tails raised and eyes and ears focused towards the source of the danger. Since they are among the tiger's favourite prey, they must be constantly vigilant.

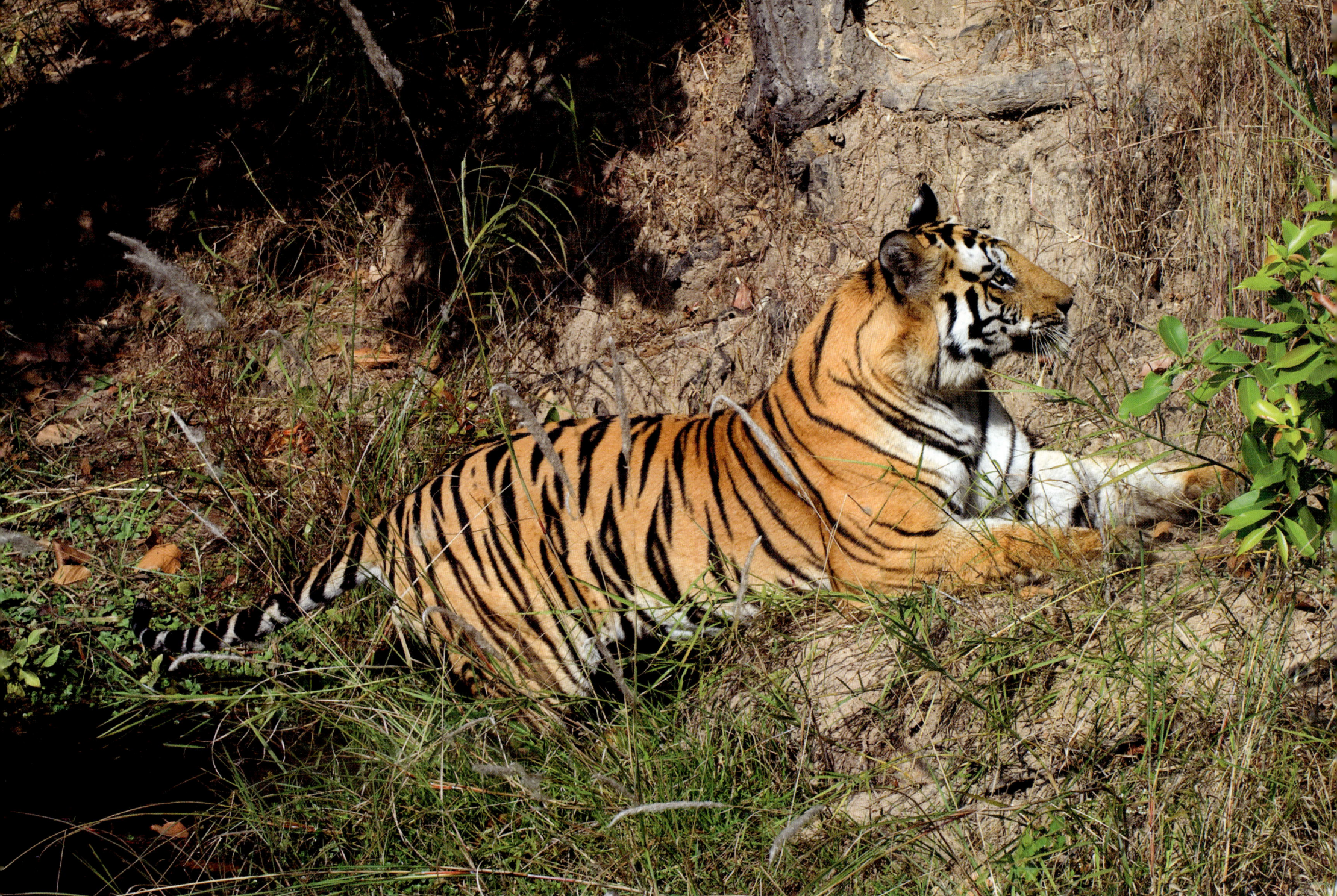

A strong tiger will sometimes even attack gaur, although these enormous wild cattle are formidable opponents. Heavy bulls can weigh up to one tonne, though females are much lighter.

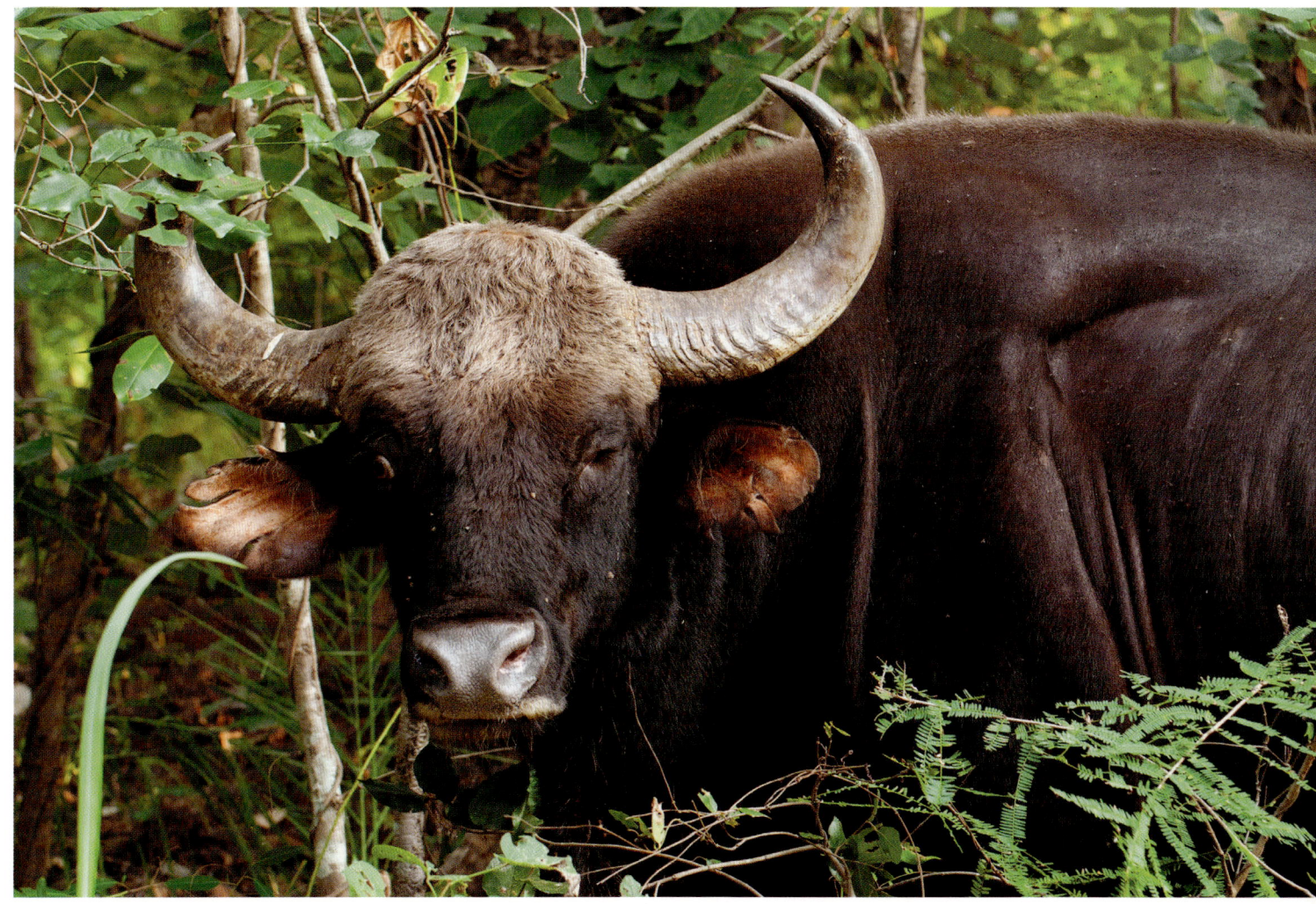

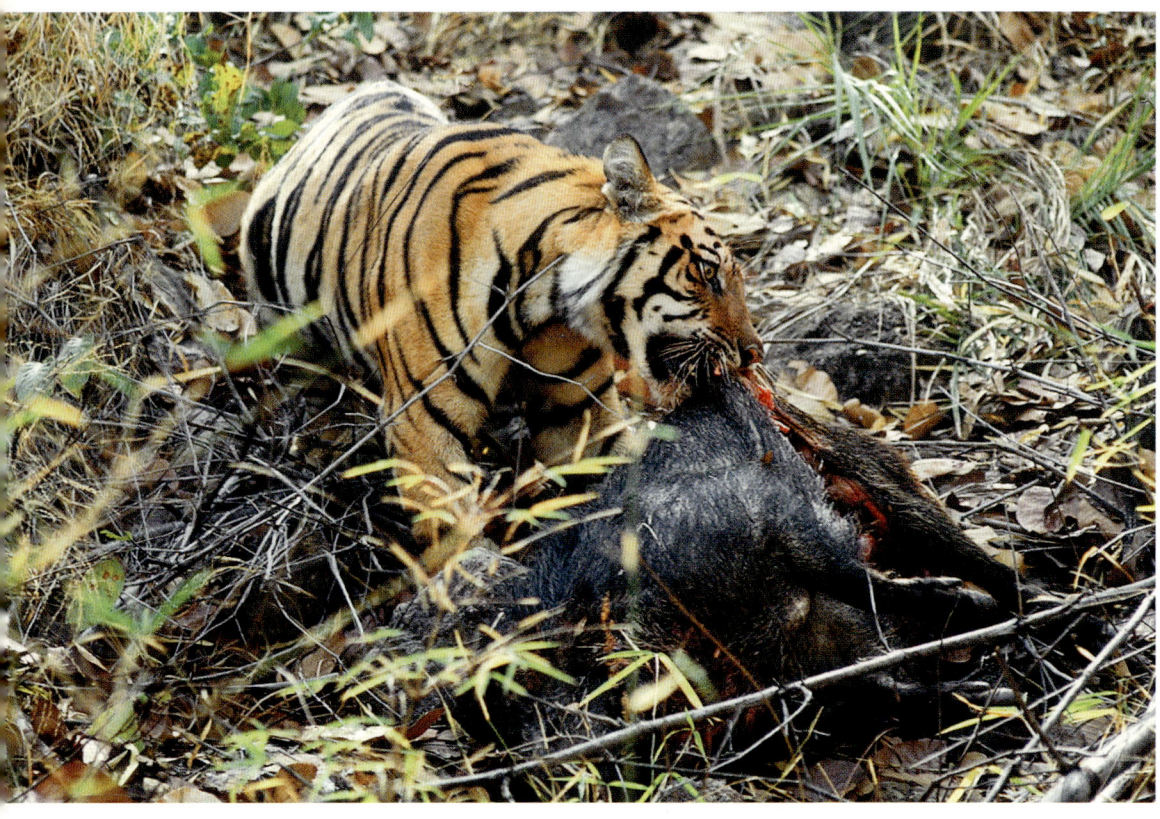

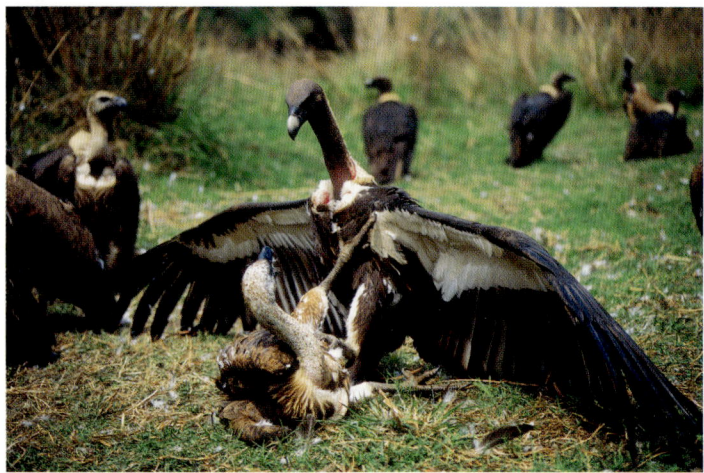

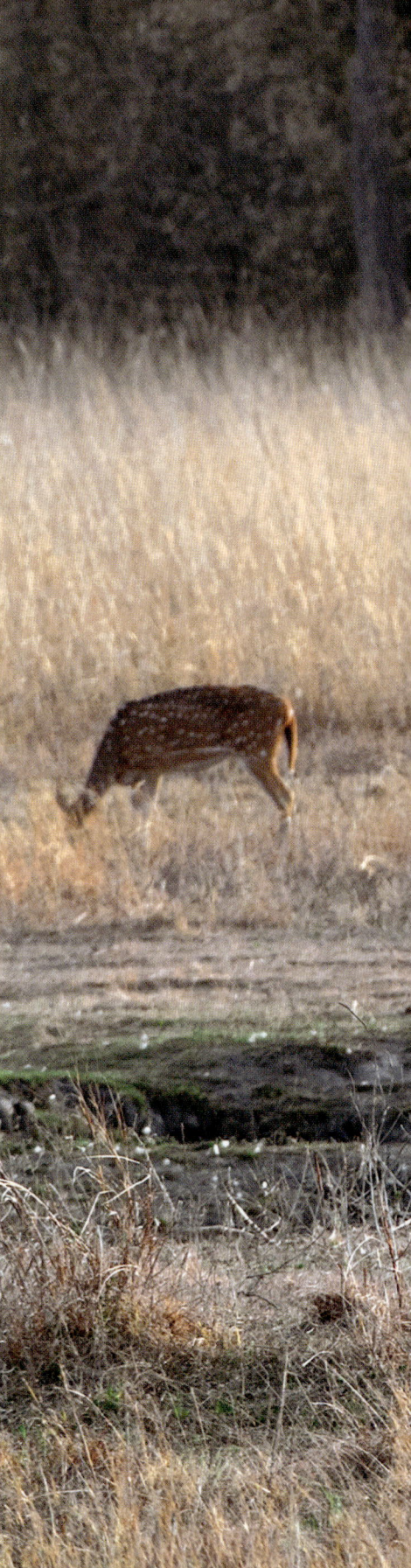

A tiger lies in ambush, watching a herd of spotted deer – in vain, for they have already seen him (right). Another tiger has had more luck, and has caught a wild boar. When white-backed vultures discover a kill, it often leads to a squabble.

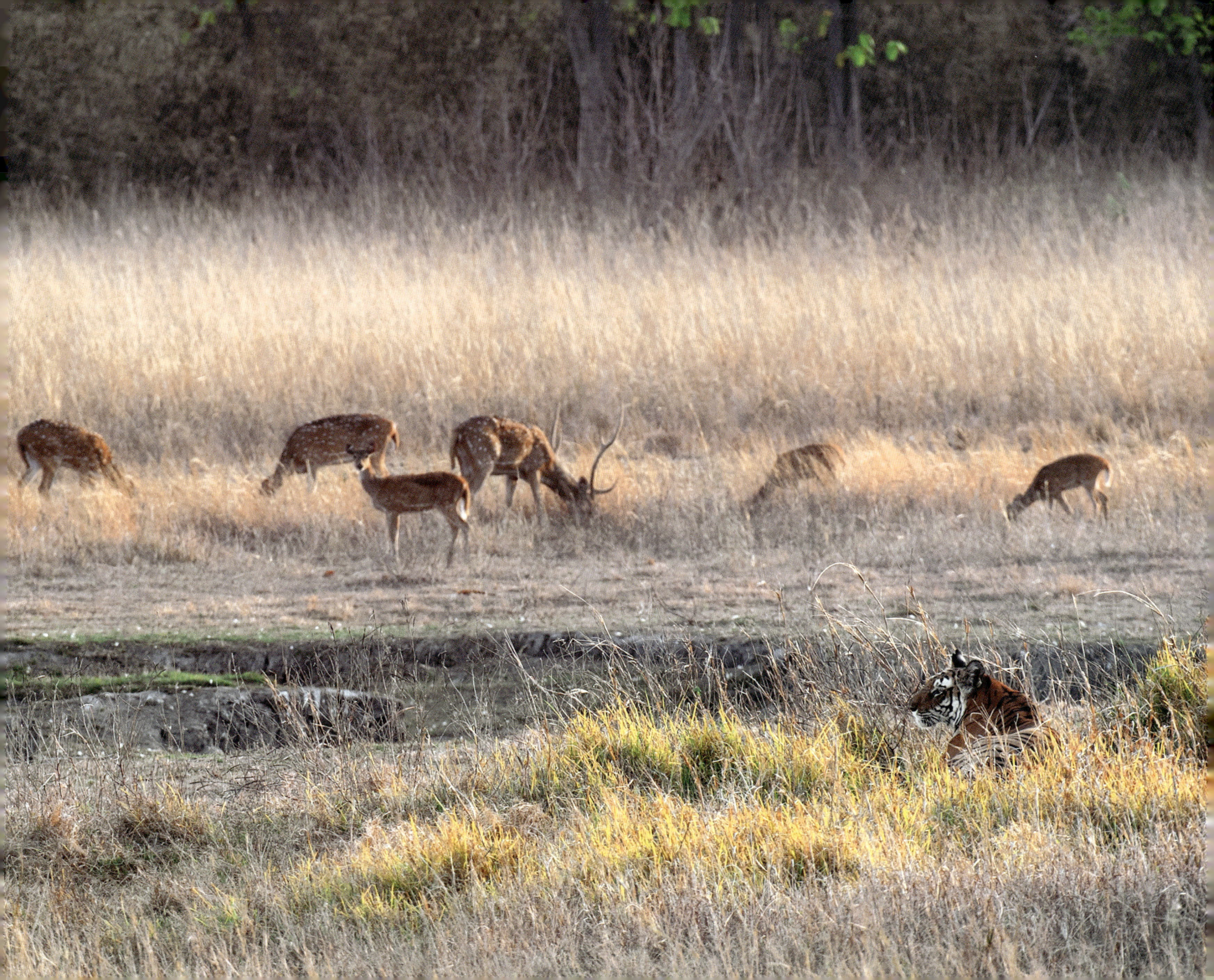

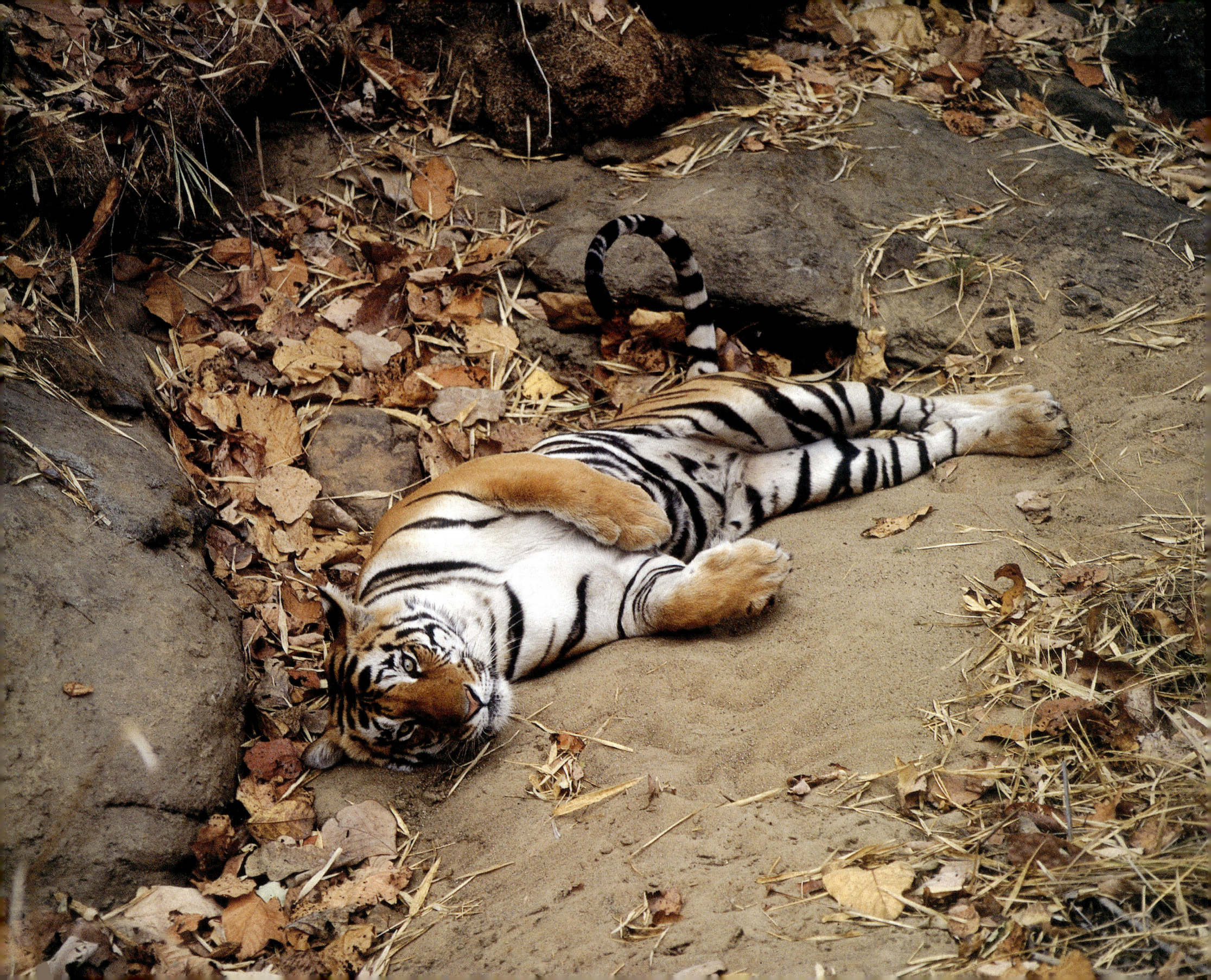

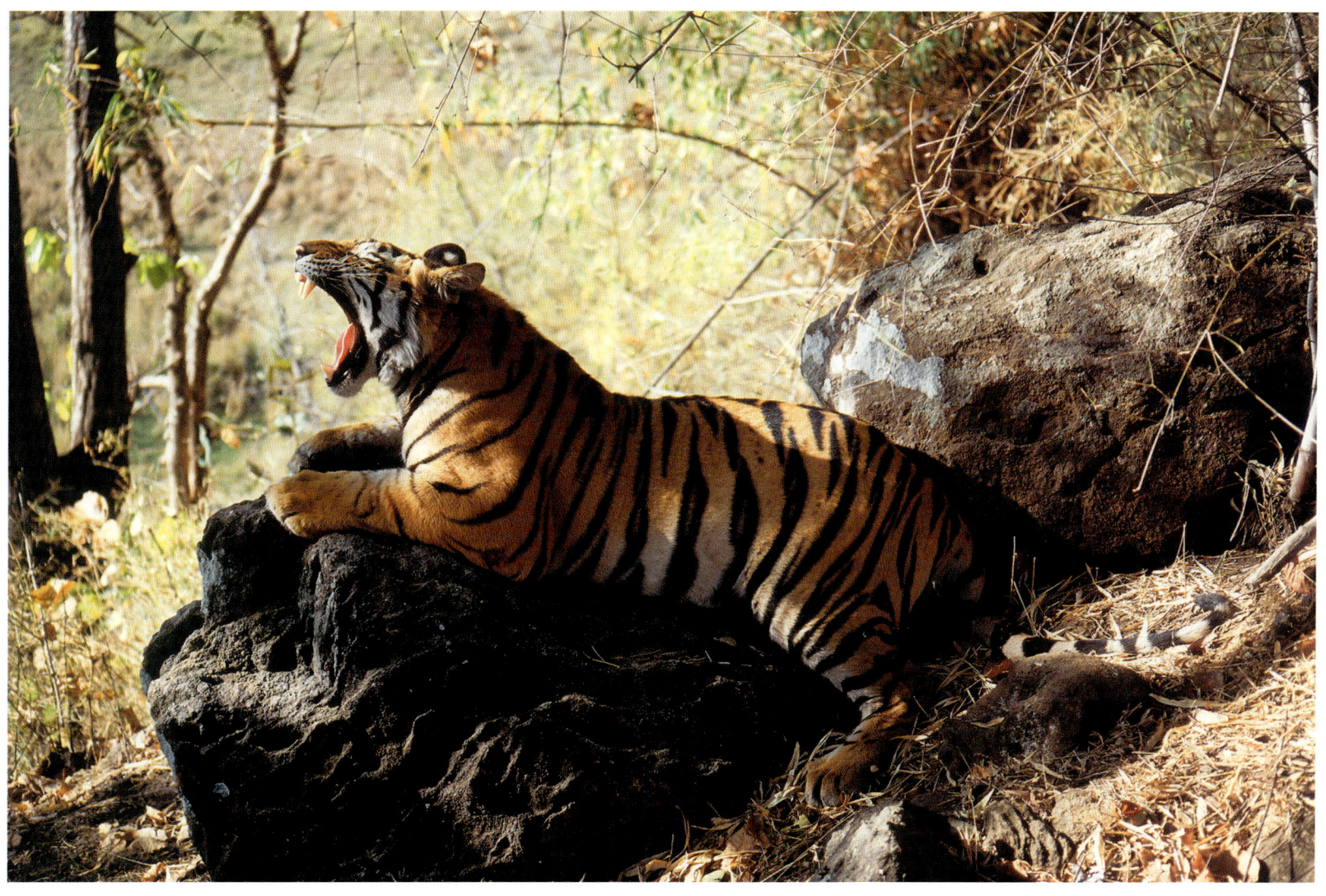

In the core areas of the national parks, where tigers have no need to fear humans, the animals often seem totally relaxed. A young tiger, raised here, is confident of his role as lord of the jungle (above). Undisturbed and well-run protected areas are essential for the survival of the species.

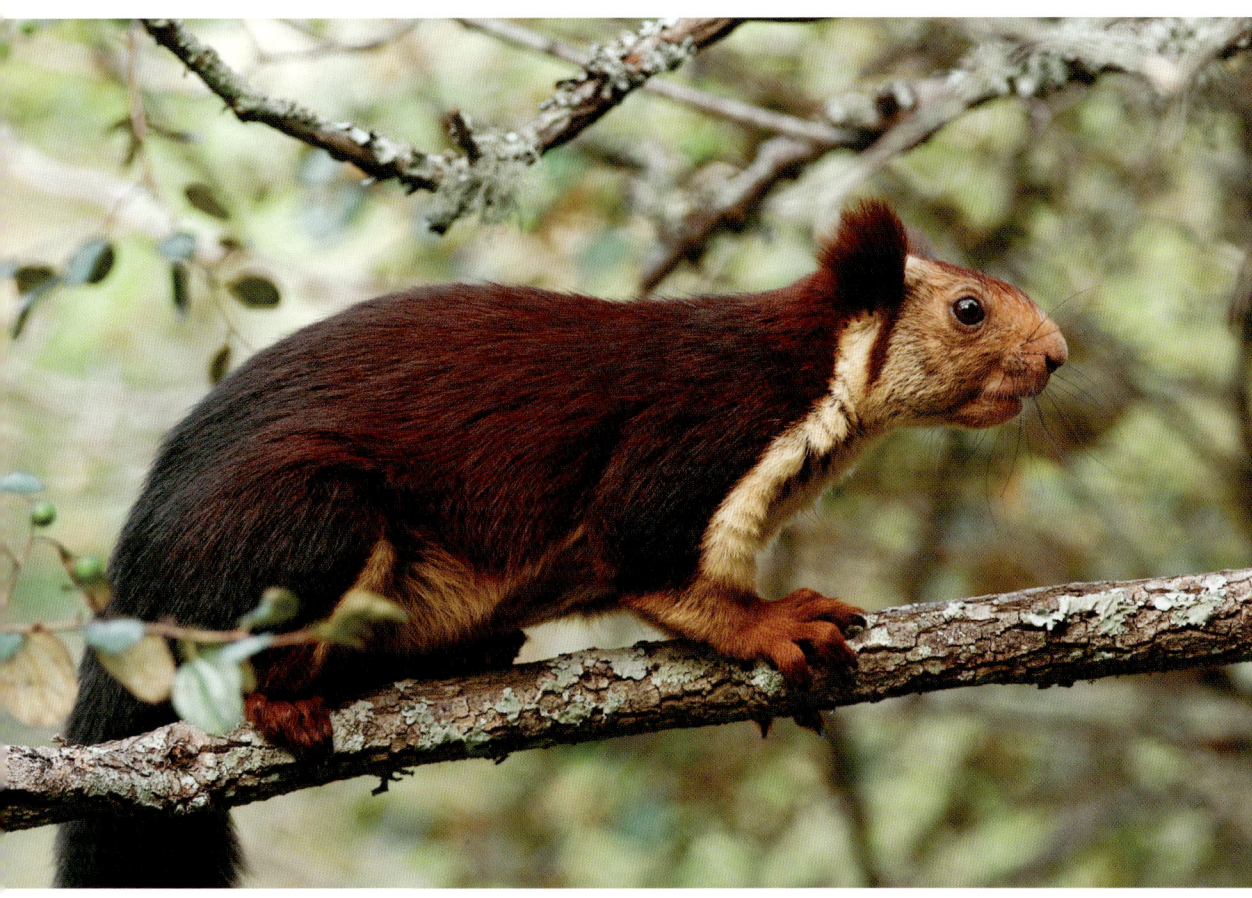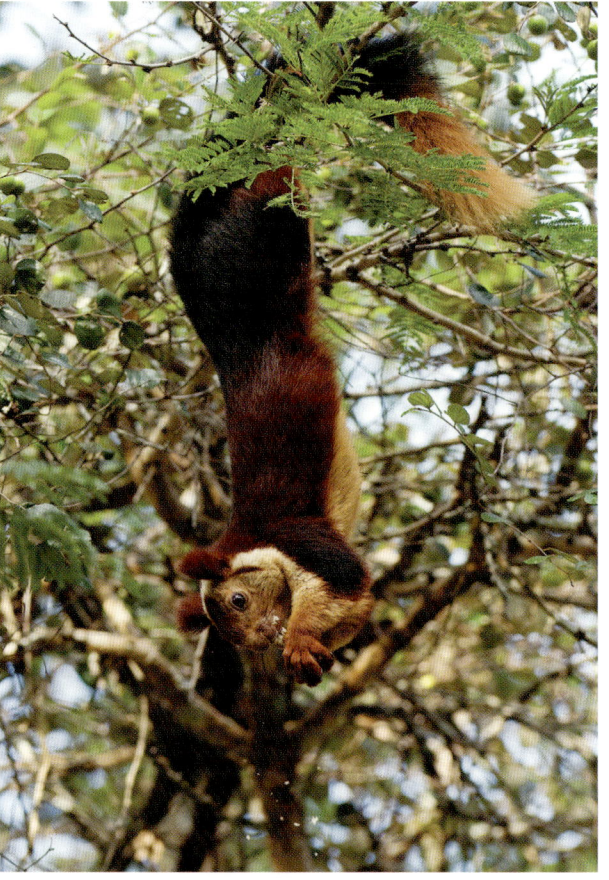

Indian giant squirrels move through the treetops with great agility. They are related to European squirrels, but are the size of domestic cats.

Many species benefit from the protection of tiger habitat. A pair of white-bellied wood-peckers checks out a tree to see if it could provide a suitable nesting site. As well as the striking red crown, the male has red moustachial stripes.

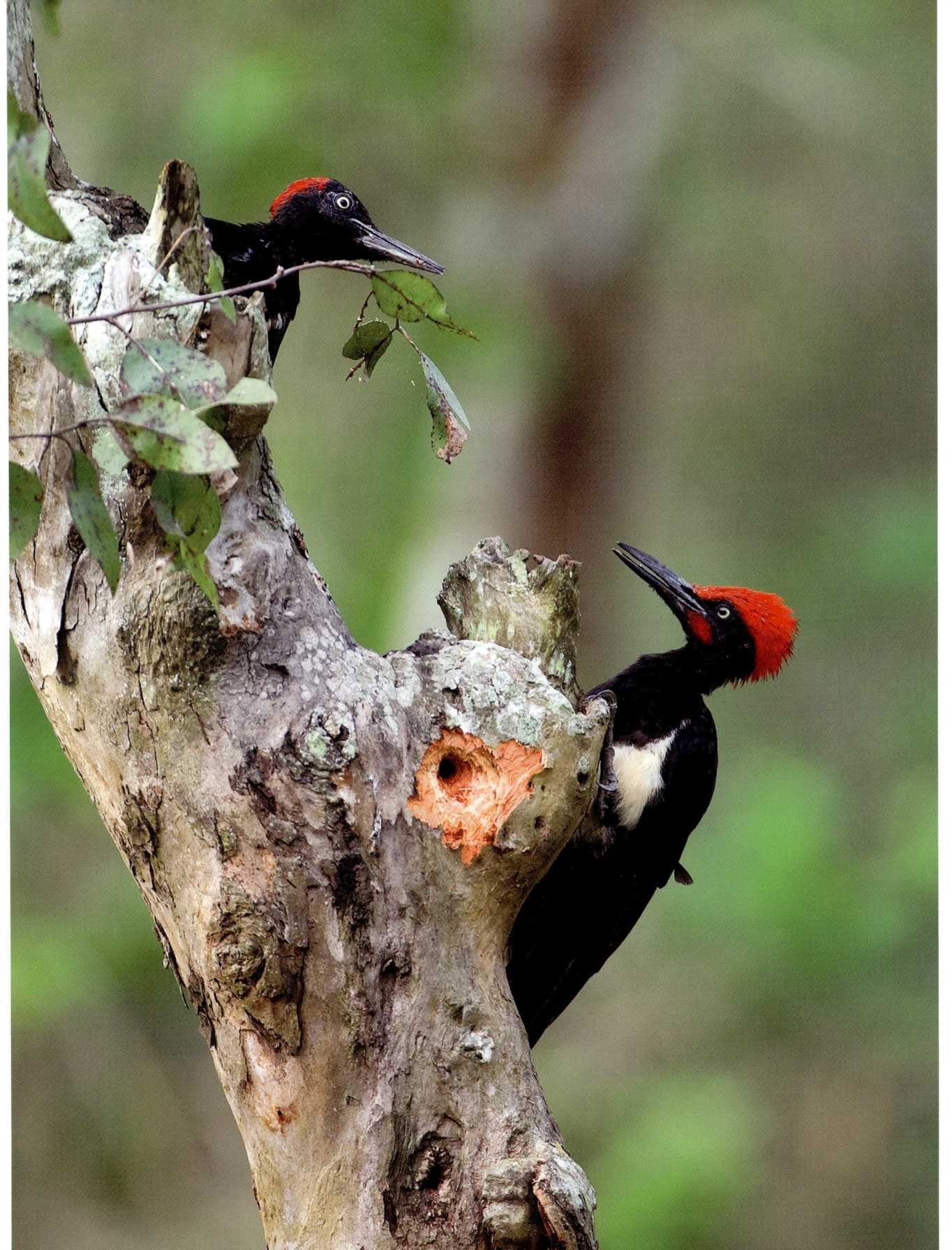

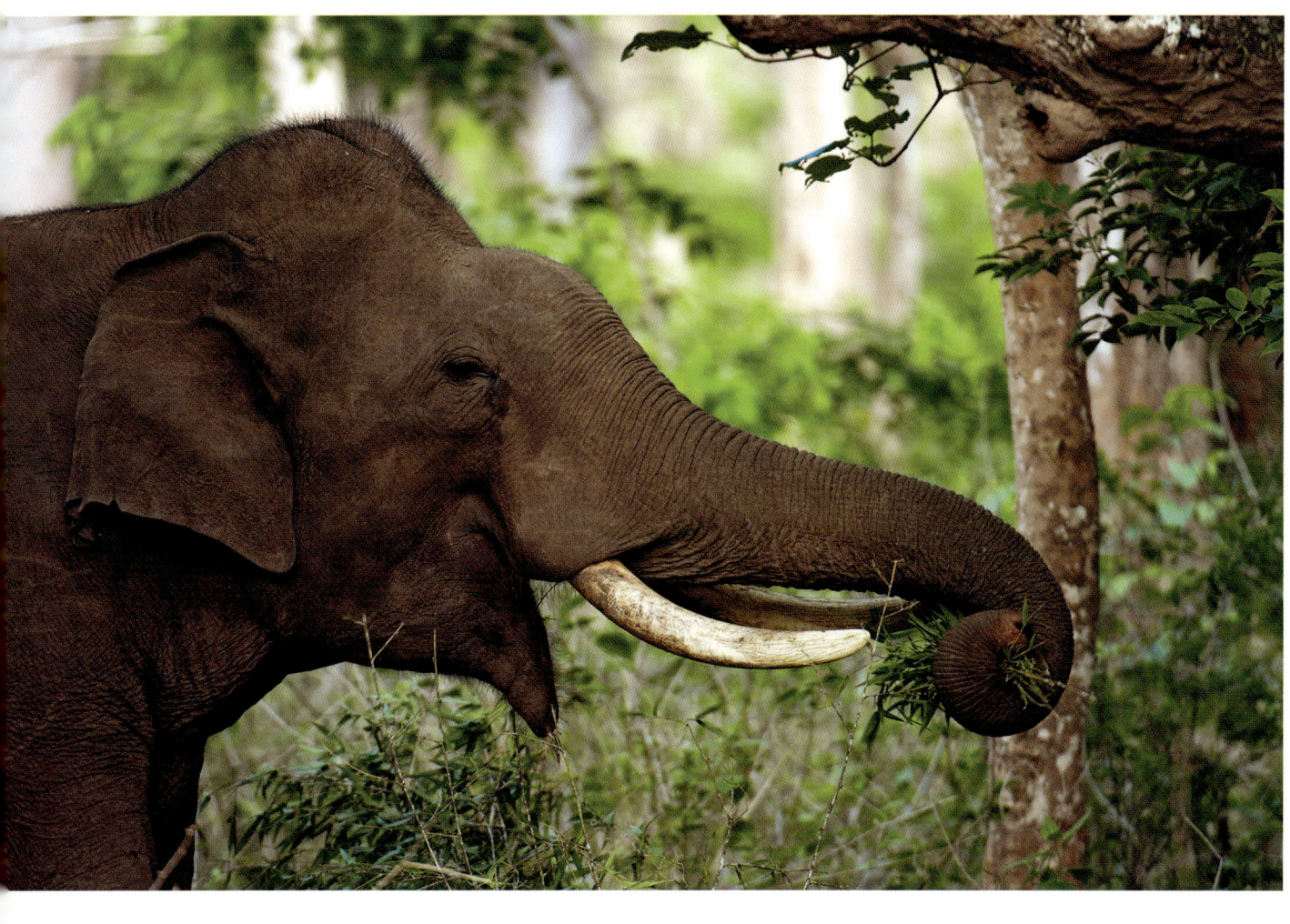

The fates of tigers and Asian elephants are closely entwined. Both species need large contiguous forests, and often they live in the same area, as here in Nagarhole National Park. Unlike the male (left), the female (right) has no tusks.

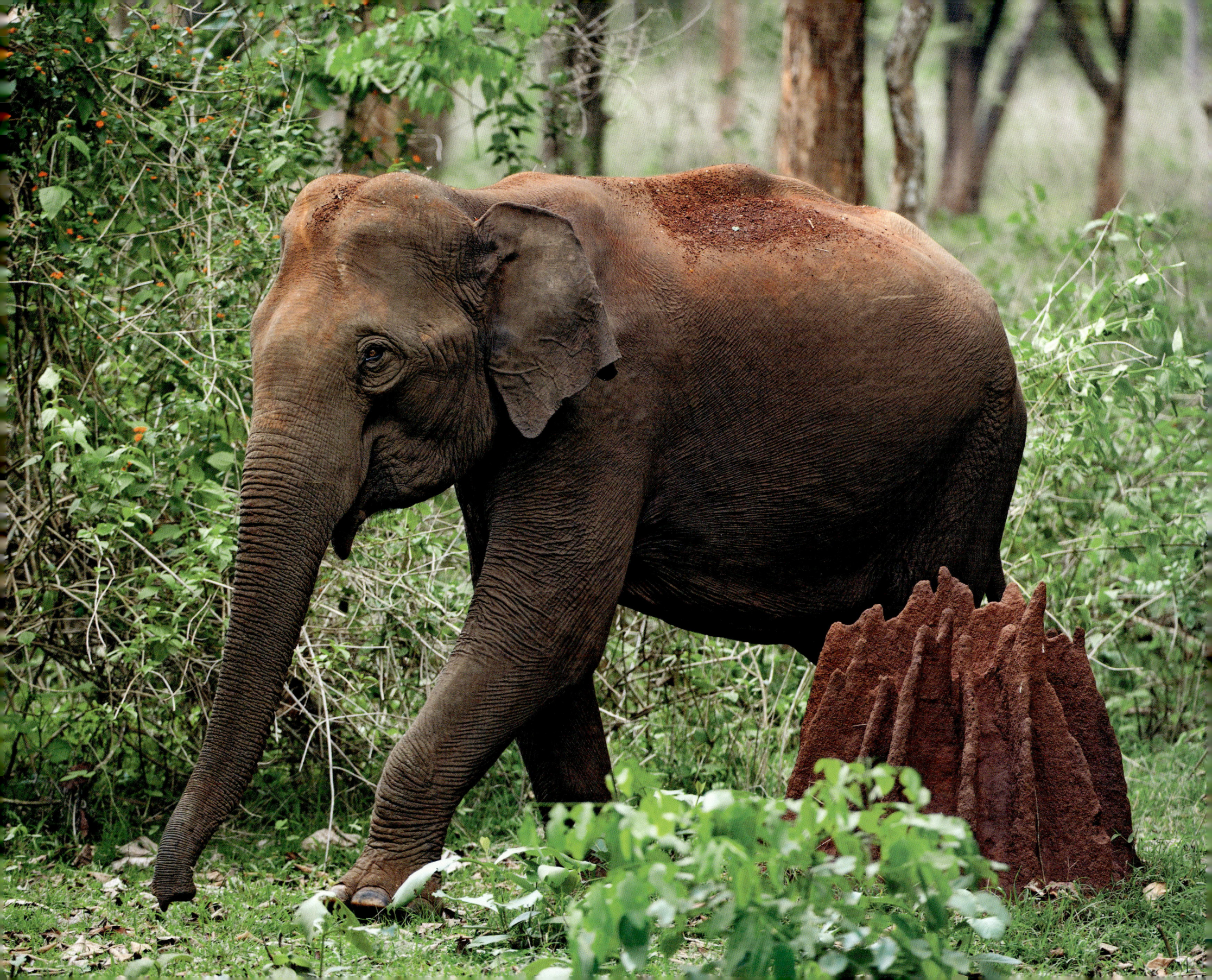

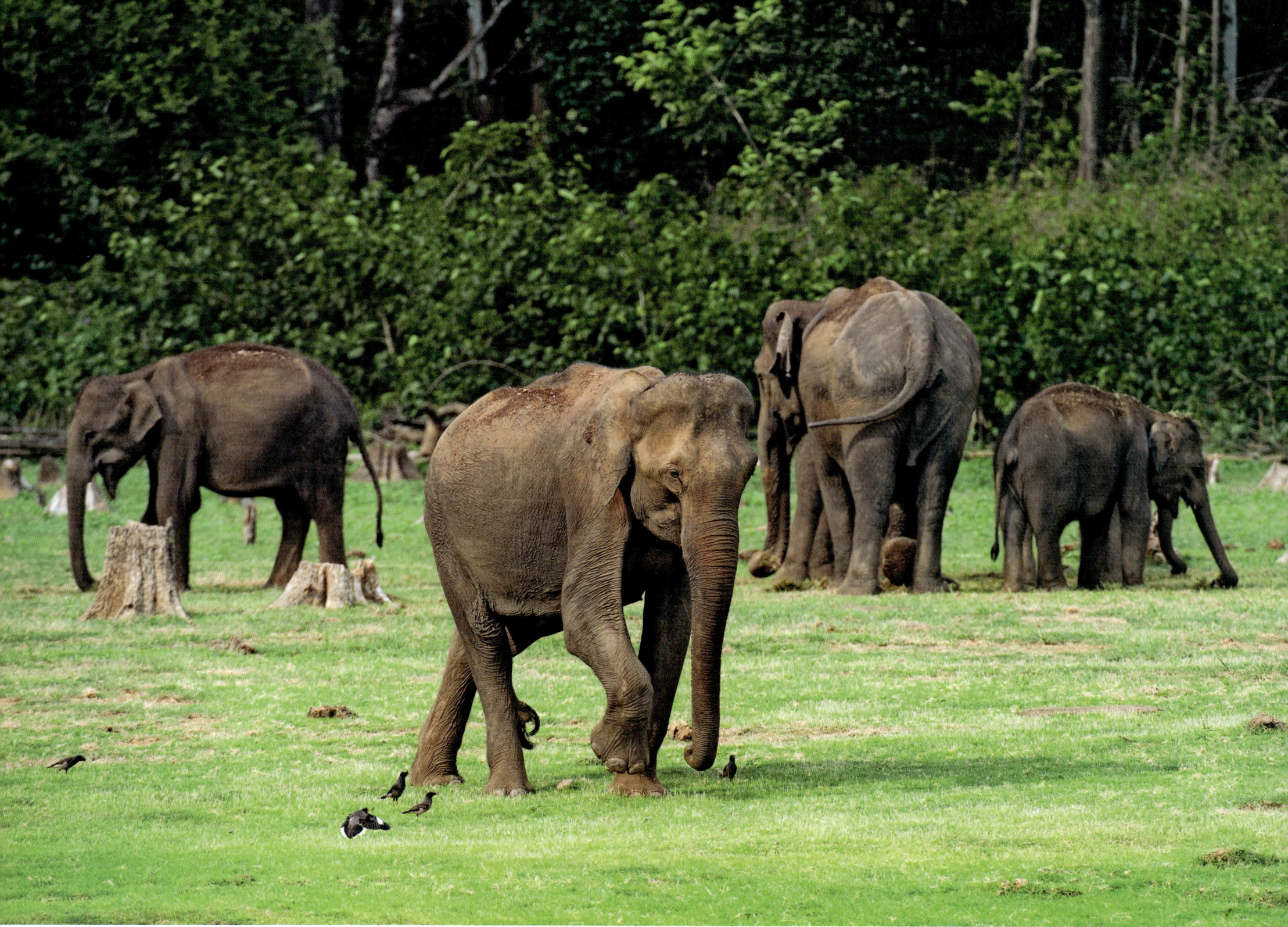

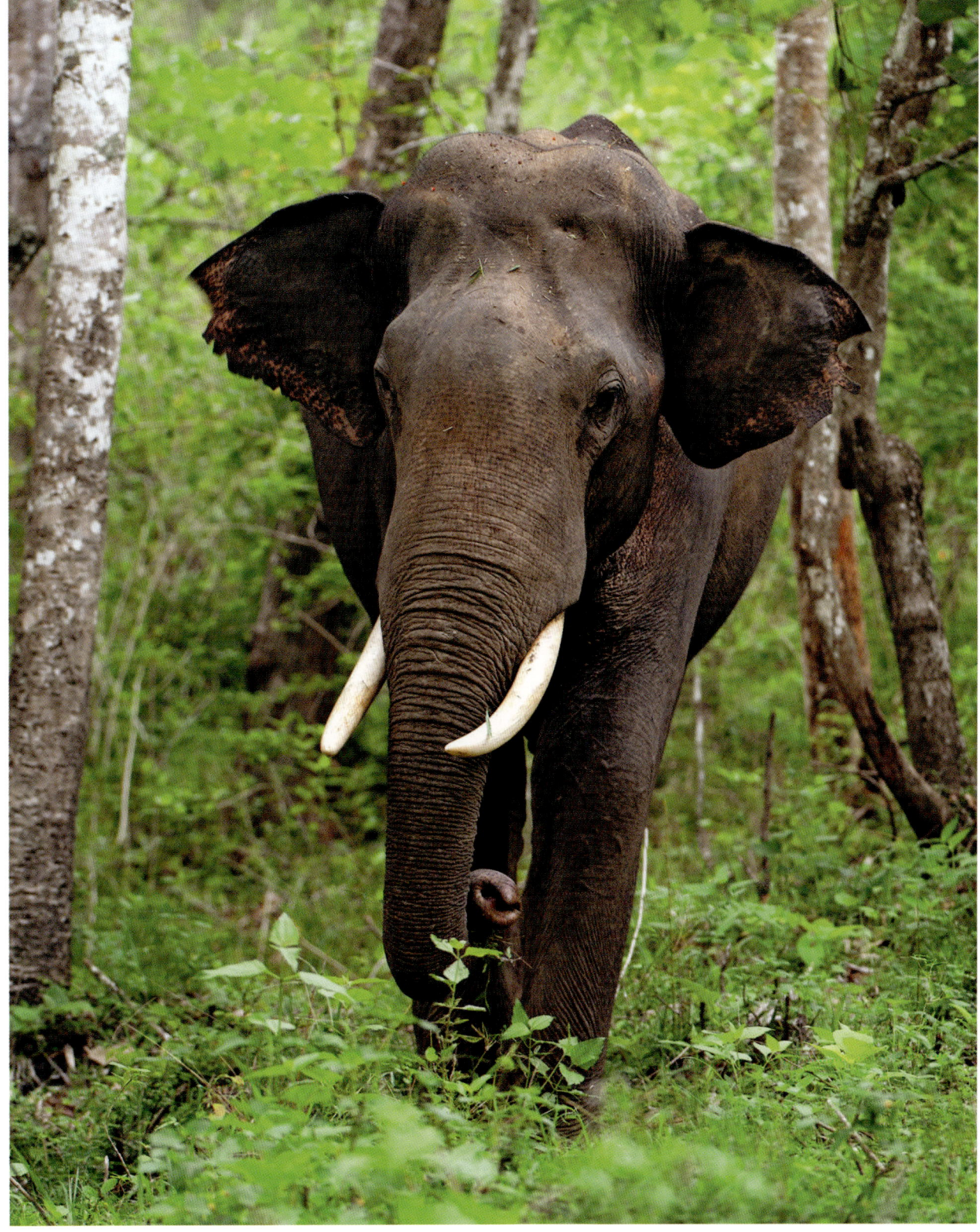

Asian elephants live in family groups led by an old and experienced female (left). Usually, all the animals in a group are closely related. Adult males have to leave the herd, and most of them roam alone through the jungle (right).

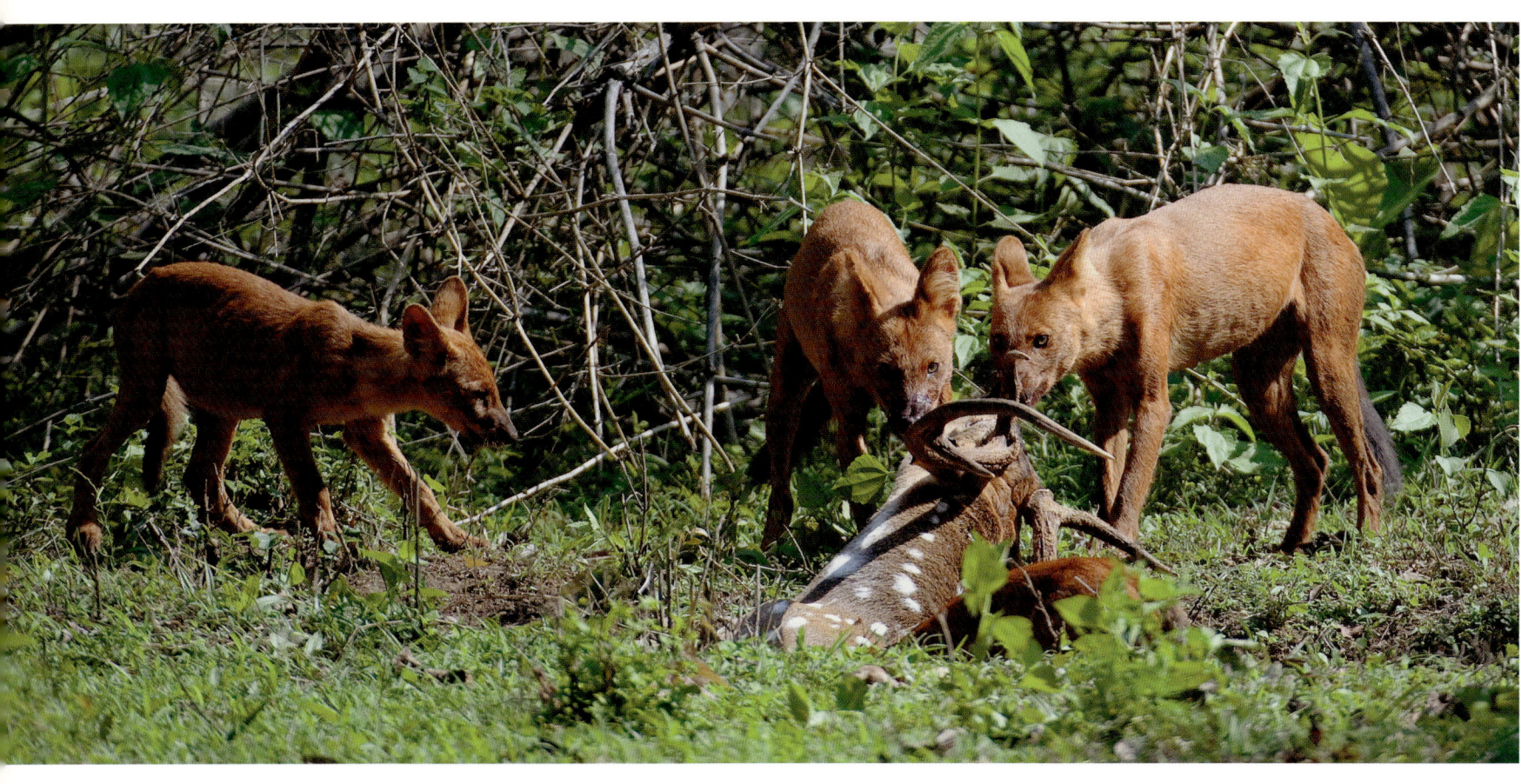

Dholes or Asiatic wild dogs are very effective hunters. Here the pack has brought down a spotted deer. At last, a pup plucks up the courage to approach the kill (above). Scientists fear that in the whole of Asia only about 2,500 dholes survive in the wild. These animals are rarer than tigers.

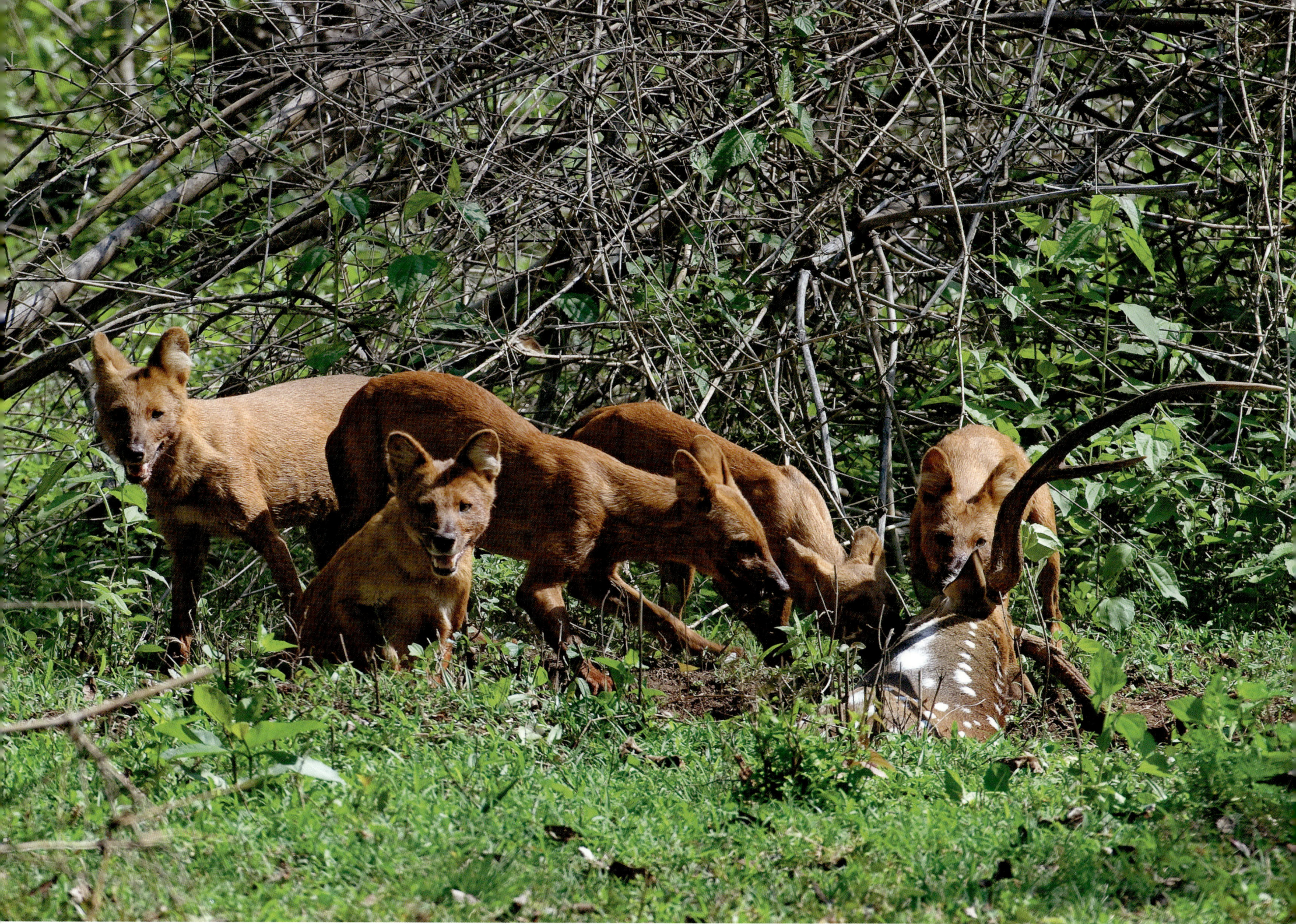

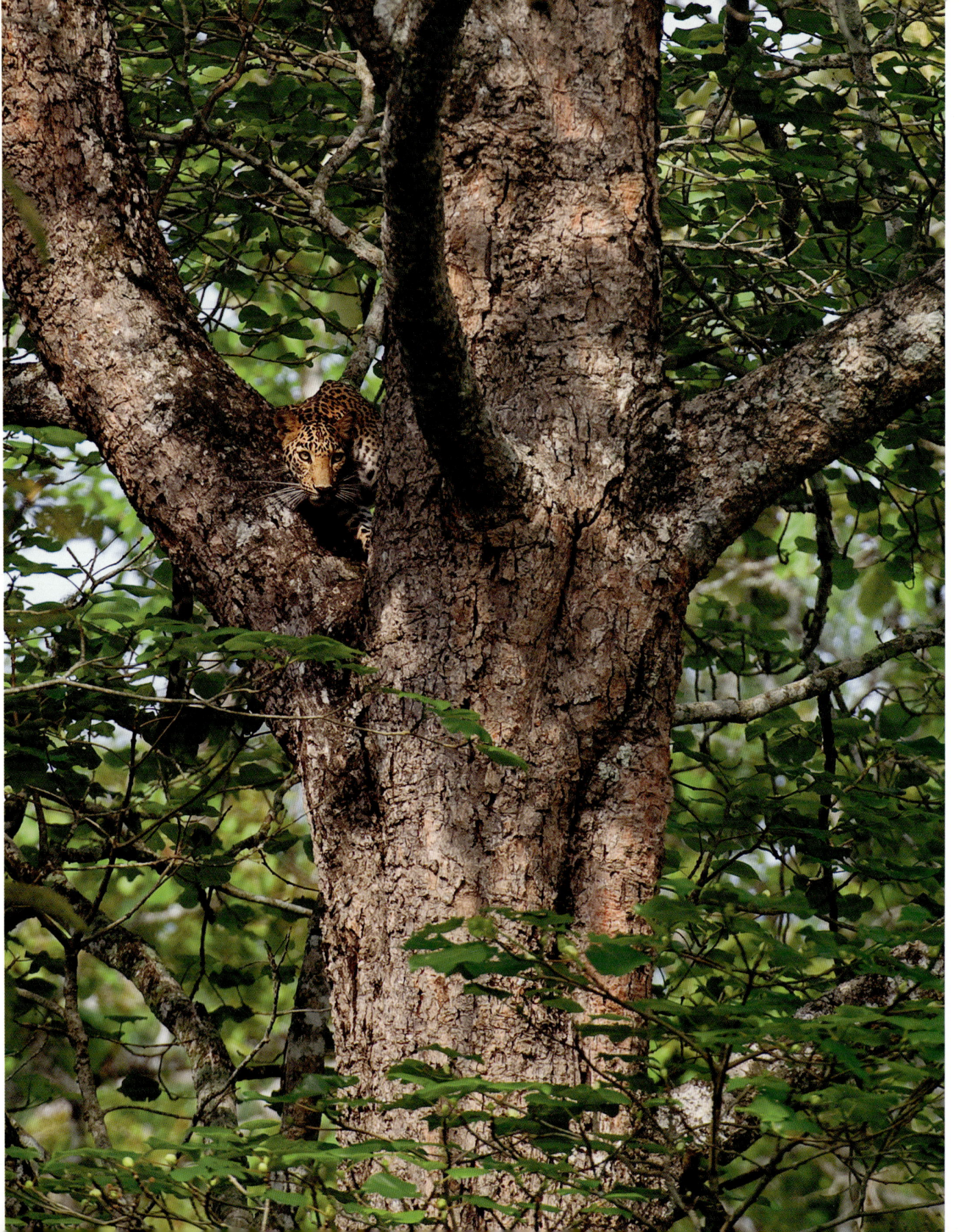

In the realm of the tiger, a leopard must always be on its guard. When it gets the chance, a tiger will kill its smaller competitor – so the leopards like to retreat to the higher trees. They are quite safe up there, for tigers, with their greater weight, are not so good at climbing.

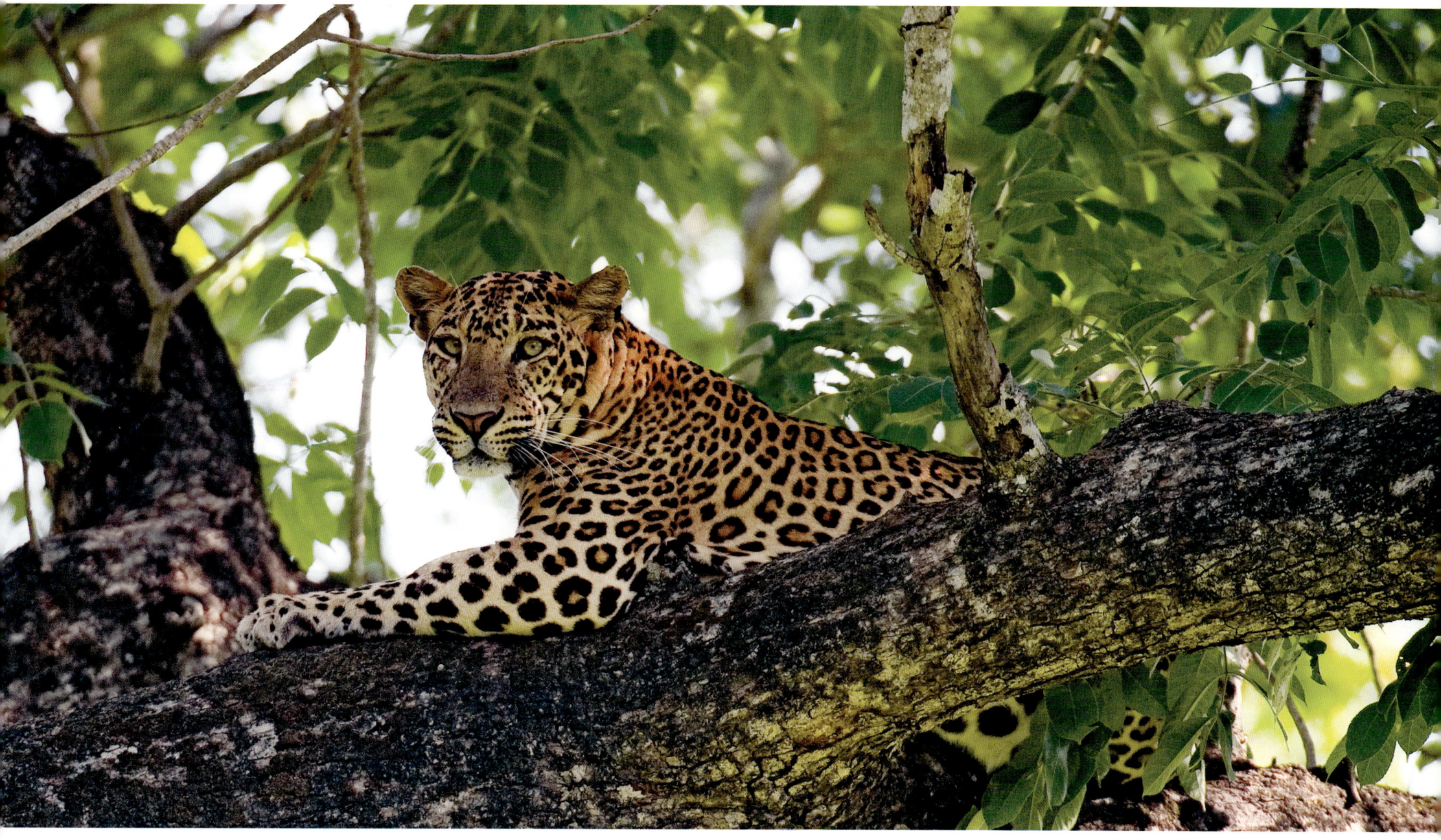

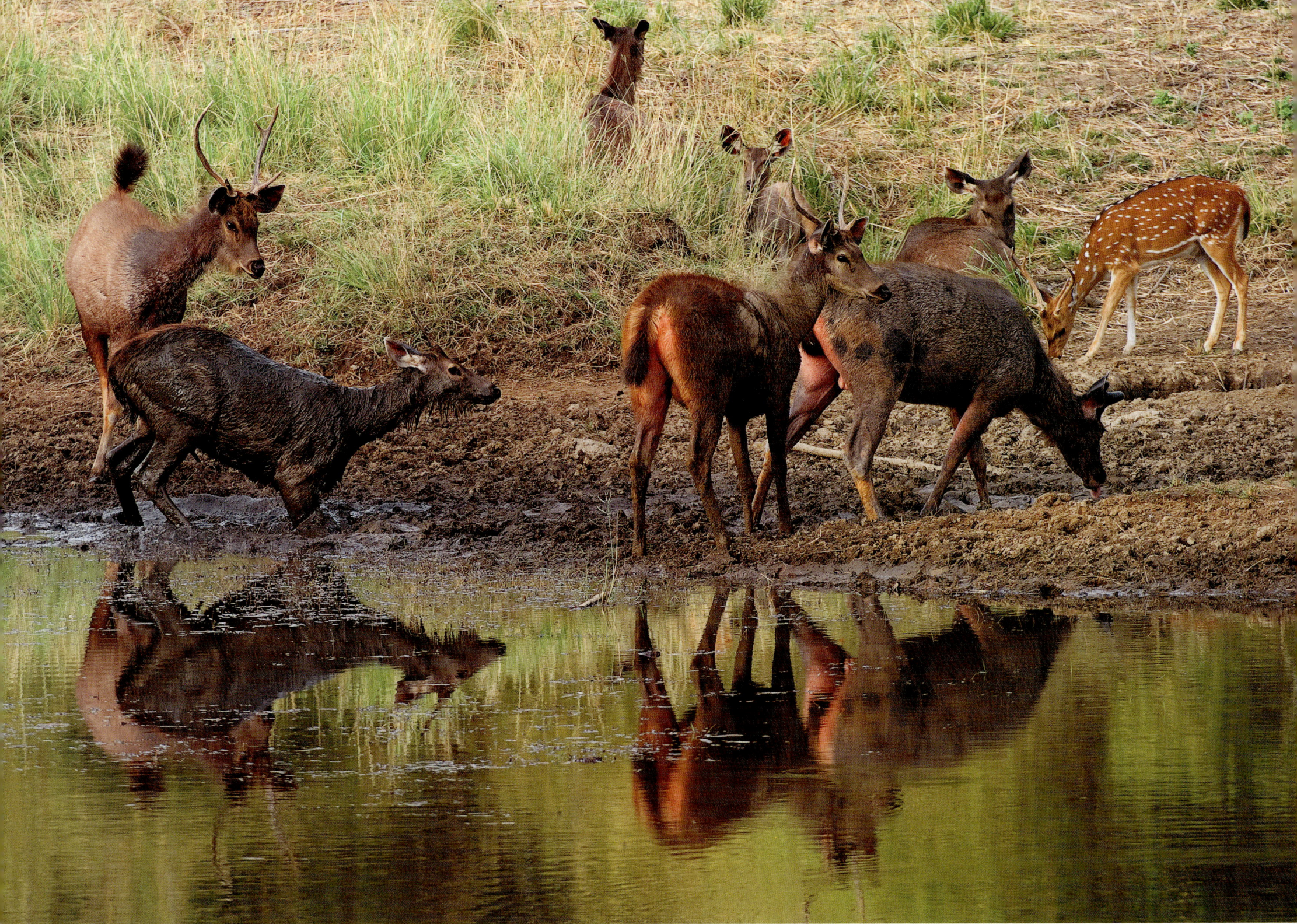

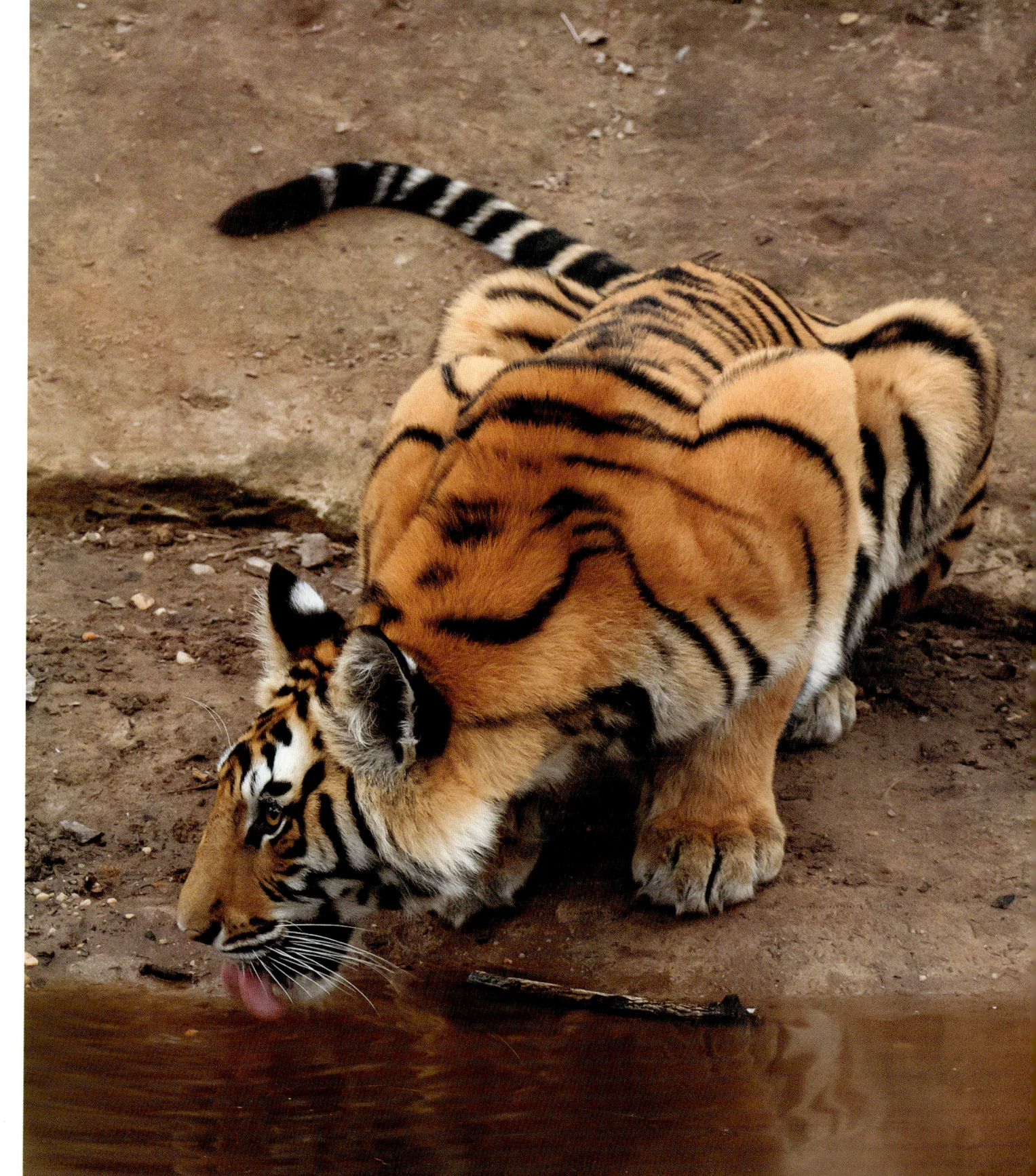

Waterholes attract predators and prey alike. Sambar deer take a mud-bath in a wallow. In the evening, a tiger comes to quench its thirst before starting to patrol its territory.

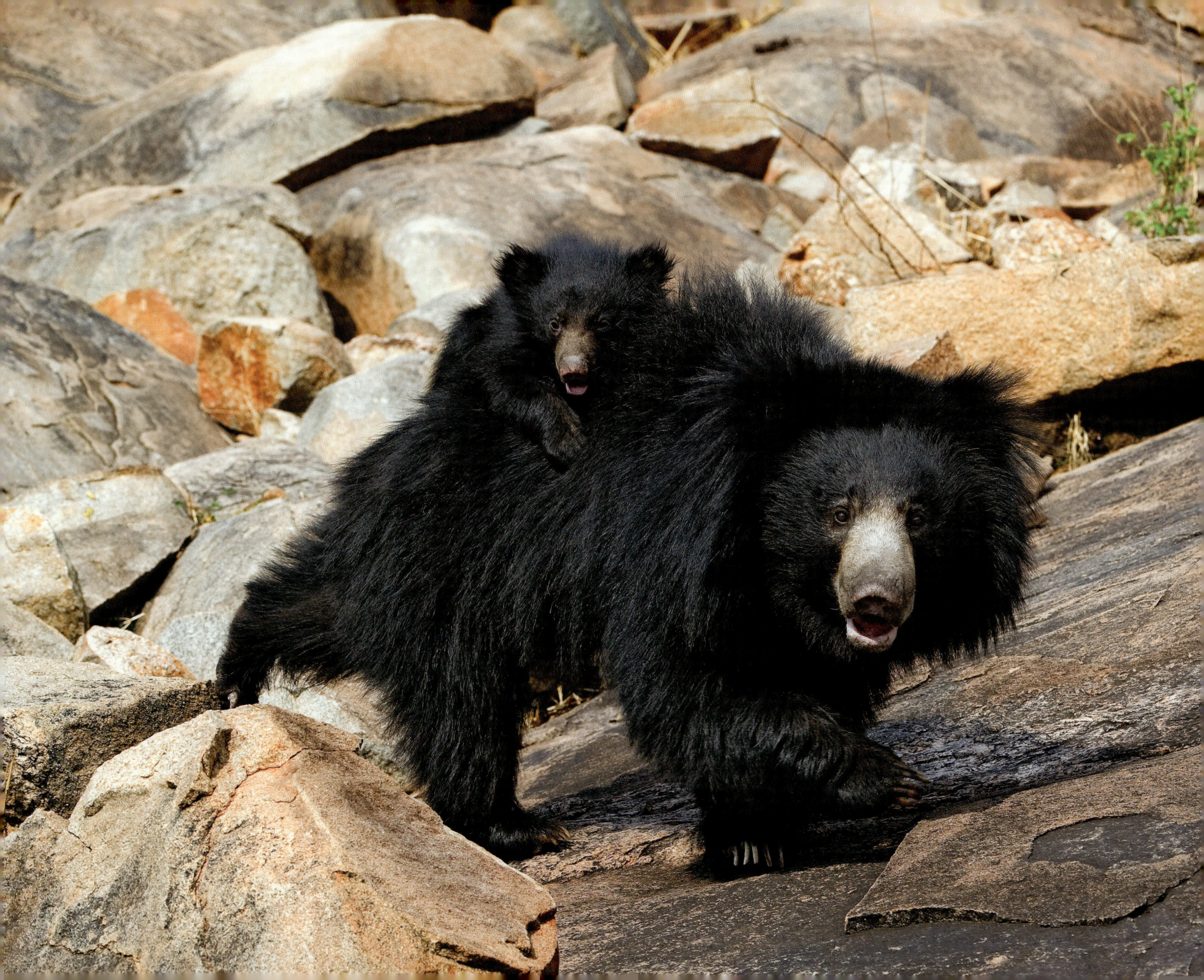

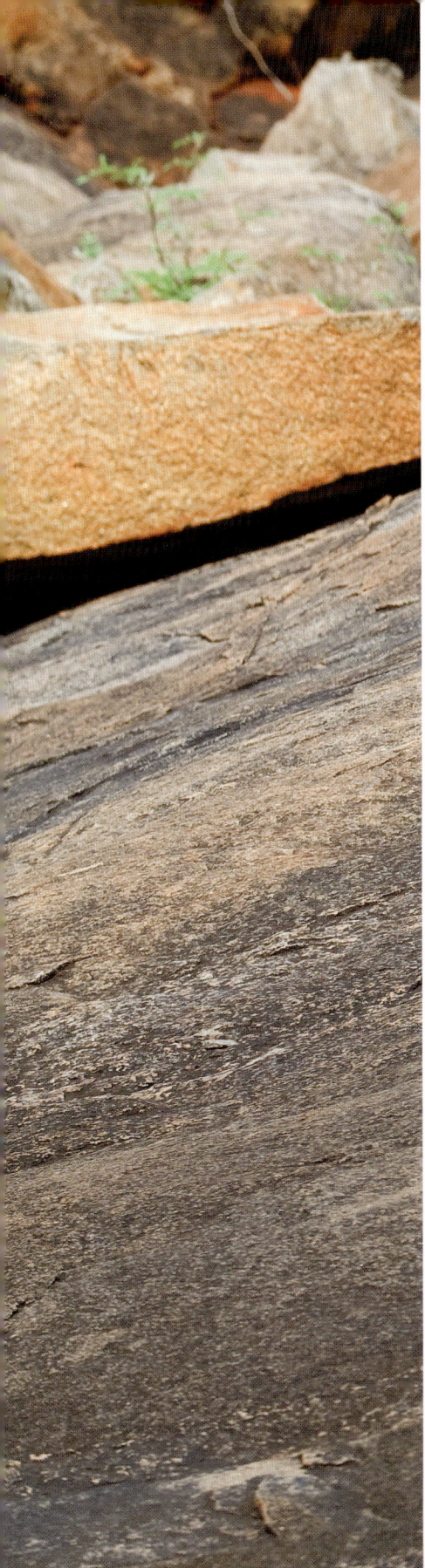

Where the Bears Roam

India

In India there lives a bear whose way of life is very little known. After a long time I finally succeeded in getting a rare glimpse of the family life of the sloth bear.

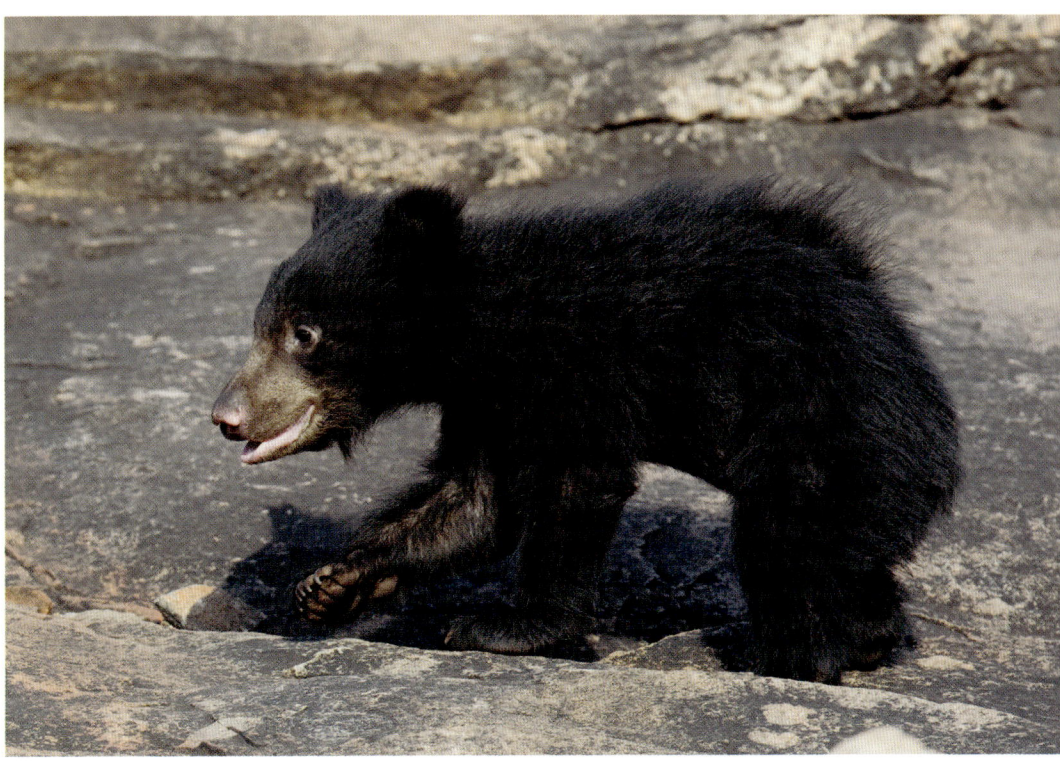

Baloo from The Jungle Book

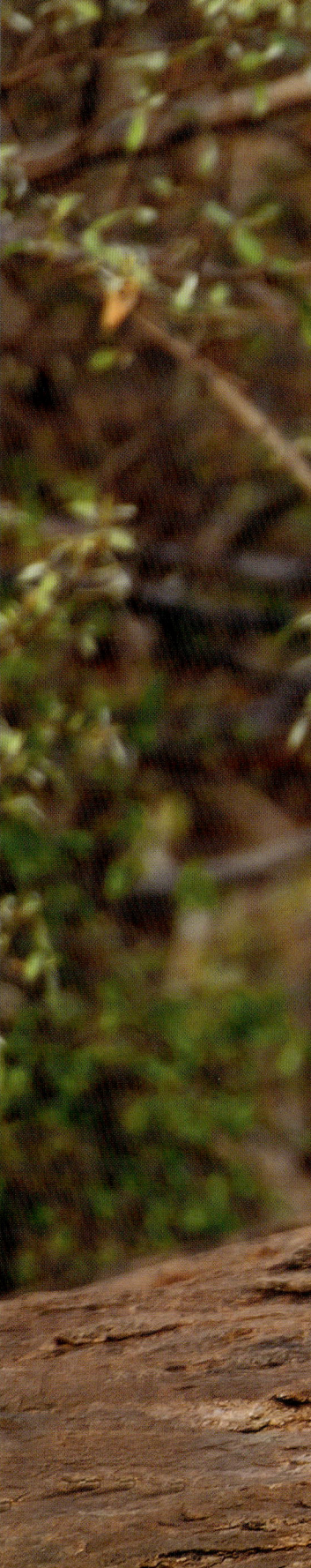

A sloth bear ambles down the hill between the cliffs. She carries her cub piggyback-fashion, and from this vantage point the youngster eagerly scrutinizes its surroundings. This is the only large bear that regularly carries its cubs on its back, an especially charming characteristic. Then the mother halts, and the little one loosens its grip on her fur, shimmies down her hind leg and begins to explore the area on foot. The mother bear searches among the stones for a tasty morsel, and the little one imitates her as best it can.

I am reminded of 'The Bare Necessities', the theme song of the Disney film *The Jungle Book*, the song that made the happy-go-lucky Baloo the Bear world-famous. The film was of course based on the writings of Rudyard Kipling, who in turn had used the animals of the Indian jungle as characters in his stories – and the real-life model for Baloo was the sloth bear. Unlike its fictional counterpart, however, this animal is known to very few people.

Even some of the basic facts about wild sloth bears are unknown to science, partly because they are overshadowed by well-known species, such as tigers and elephants, but also because it is difficult to conduct field studies on these shy and generally nocturnal bears. Remarkably, even in the 21st century, there are large, charismatic mammals that have scarcely been investigated.

The animals were long known only as dancing bears, tormented creatures that would amuse passers-by for a few rupees. These were also the first sloth bears that I set eyes on in India – the year was 1993. And then in 1997 I saw a wild sloth bear for the first time. It was a chance encounter, and I was delighted – even though I wasn't able to take a photograph. In the following years I spent a lot of time in the Indian jungle, and I saw many tigers, but only rarely did a sloth bear cross my path. I managed to take not a single picture – and I understood why there were virtually no photos of sloth bears in the wild.

I was all the more excited, then, to find myself in bear territory in the southern Indian state of Karnataka – a bizarre, rocky landscape with numerous caves. I had heard that the area was an excellent place to observe the animals. A mother bear lived here with her cubs. The mothers have greater energy requirements, because they need to produce milk for their young, and so they are more commonly active during the daytime searching for food. I took up my position in a hiding place with an excellent view and waited for my opportunity.

On that first visit, in 2007, I was lucky. I was able to observe and photograph the mother and her clumsy cubs – much better than I had dared to hope. In the following years I returned many times and saw several different bears. There were other females with cubs of different ages, lone males and young bears, already separated from their mother but still sticking together. Over time I became practised at distinguishing the individuals, so that I was able to make detailed observations on the interactions between females, cubs and males. Some of my observations and photographs even provided data on behaviours that were previously unknown in sloth bears, such as aspects of the unfamiliar social life of these animals. I was fortunate enough to put together the first photo story on wild sloth bears. It was published in numerous magazines worldwide – I was overjoyed! Sometimes it is still possible, even today, to enter uncharted terrain.

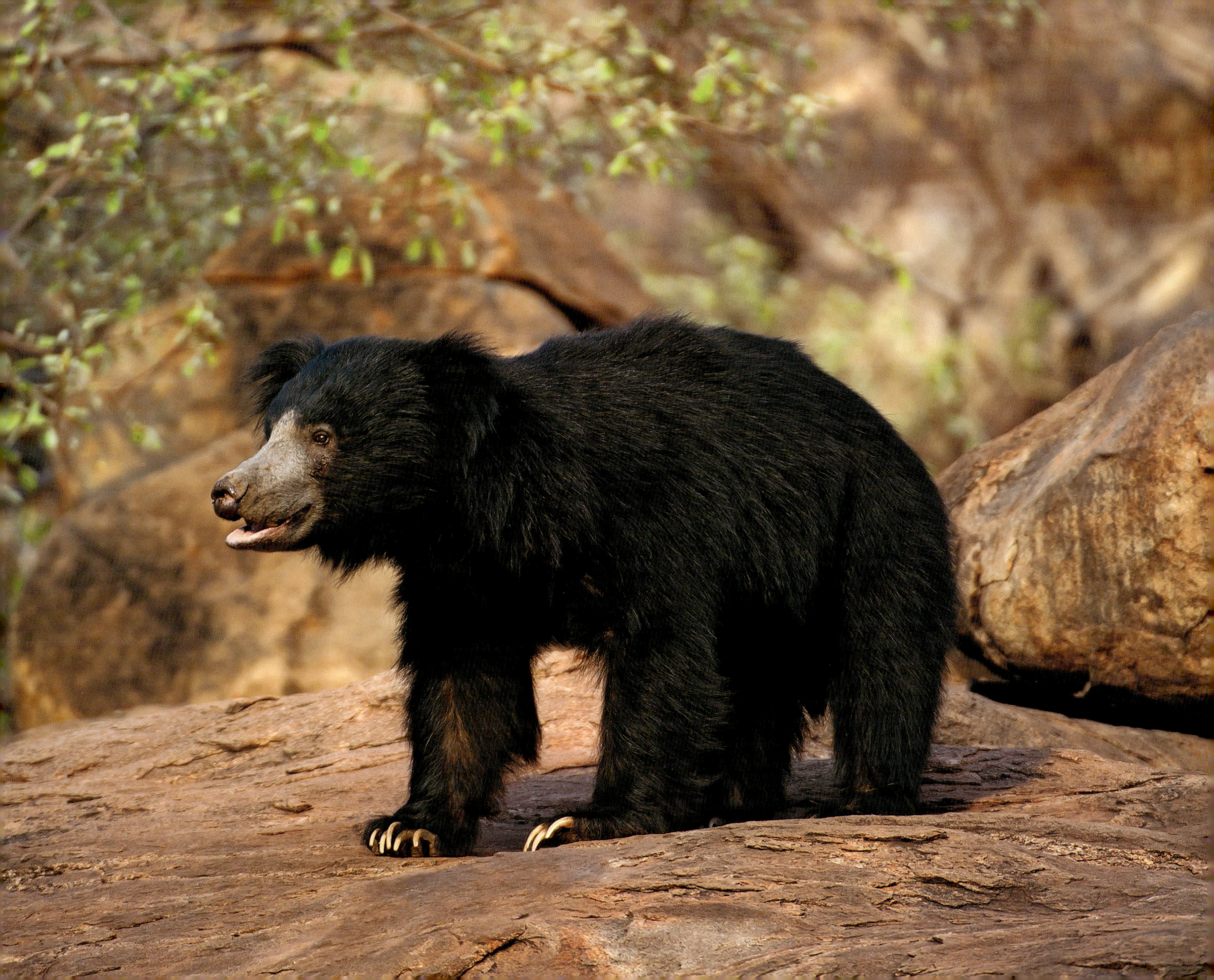

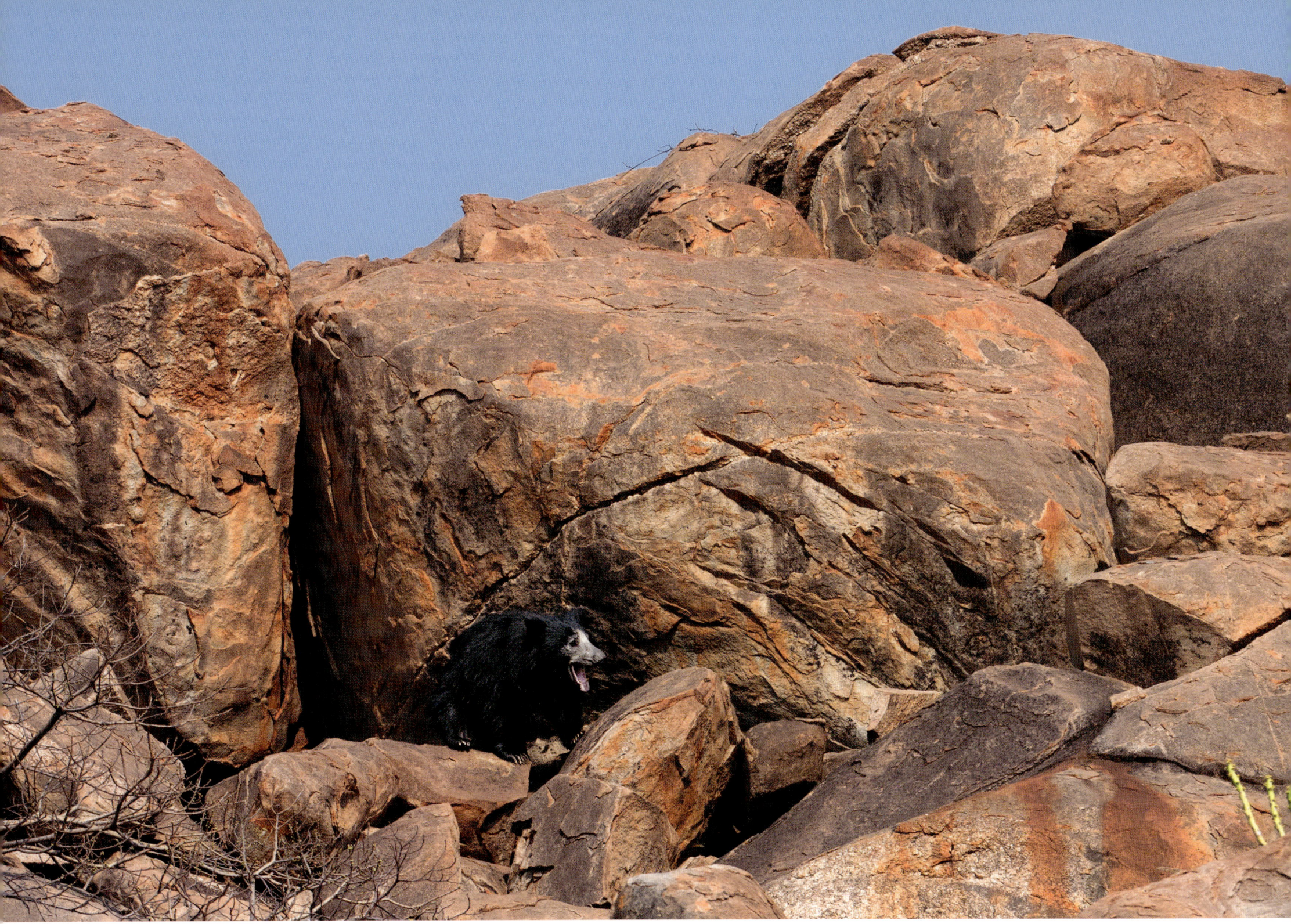

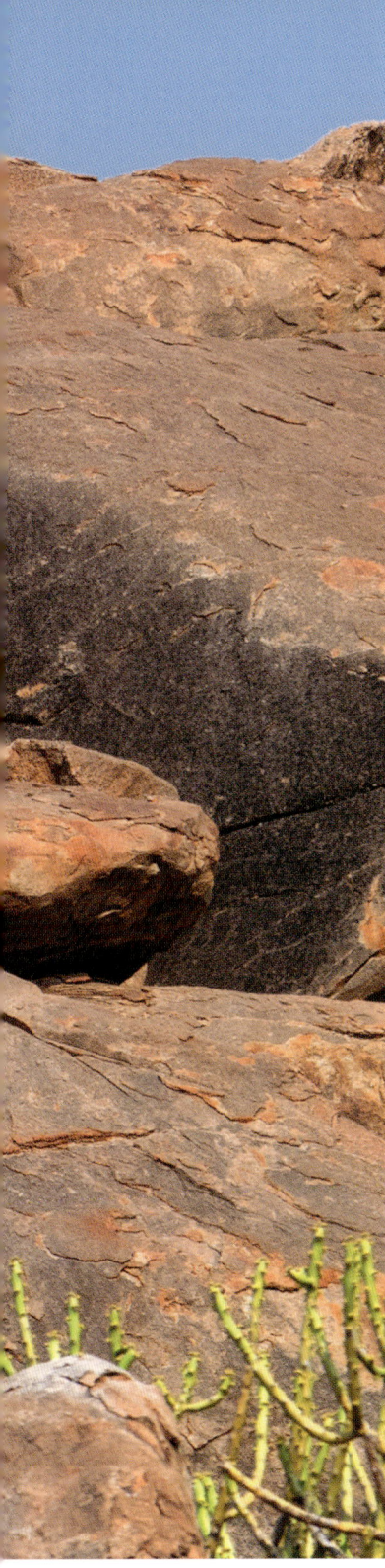

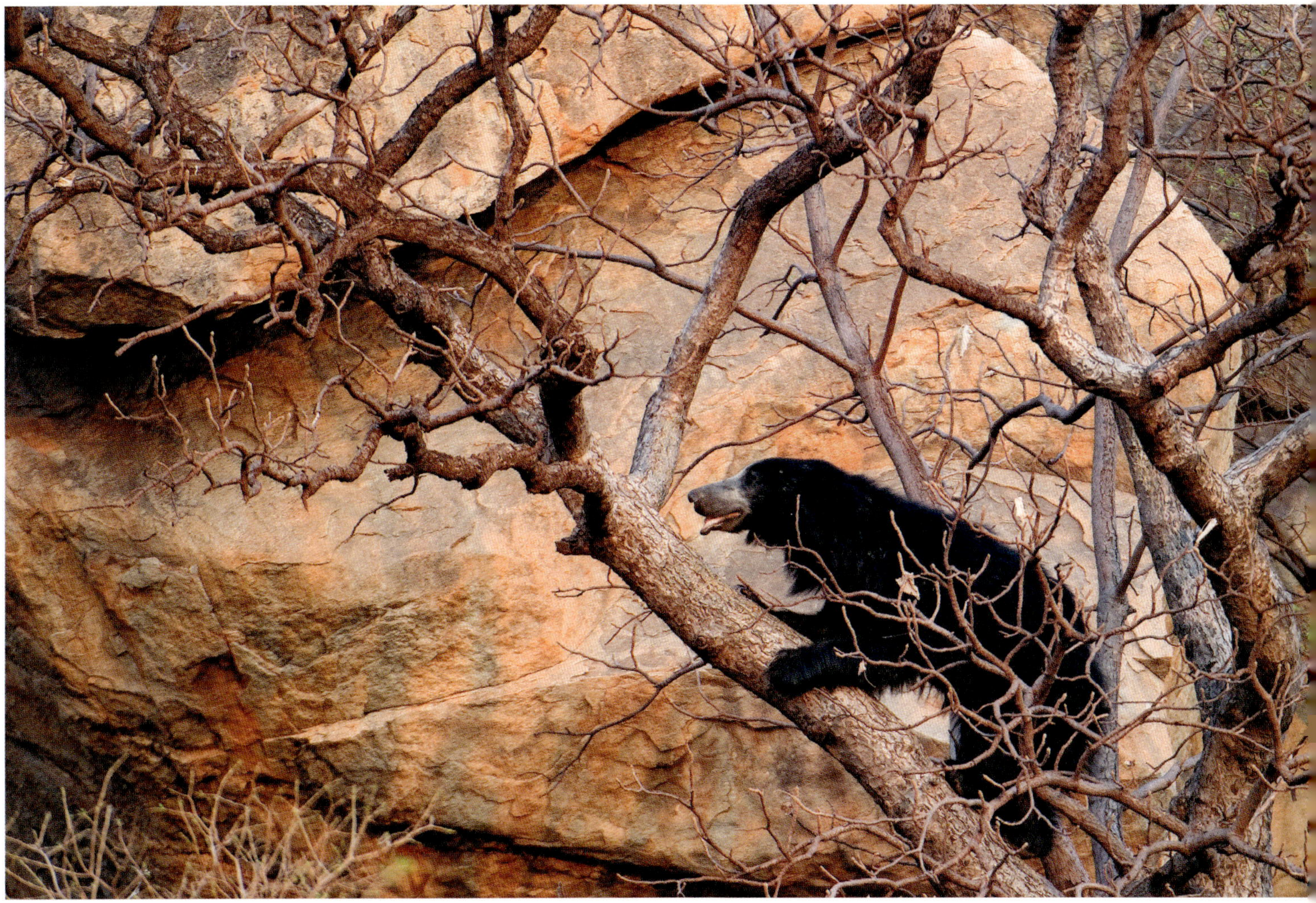

A sloth bear emerges from its lair in the rocky hills of Karnataka. These animals mainly feed on insects, such as termites and ants. Depending on the season, they also take a lot of fruit. The bears are nimble climbers, and are sometimes seen foraging in trees.

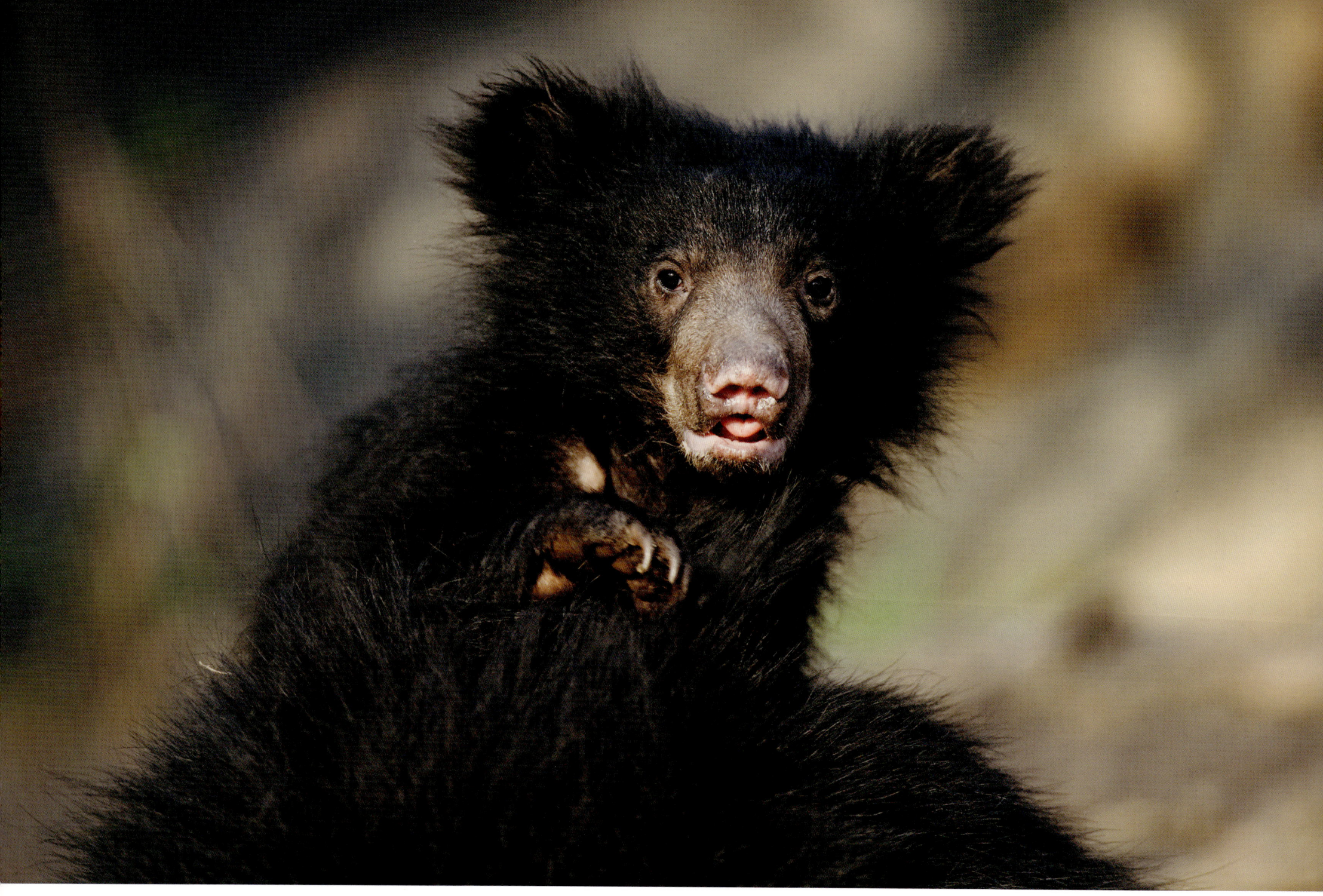

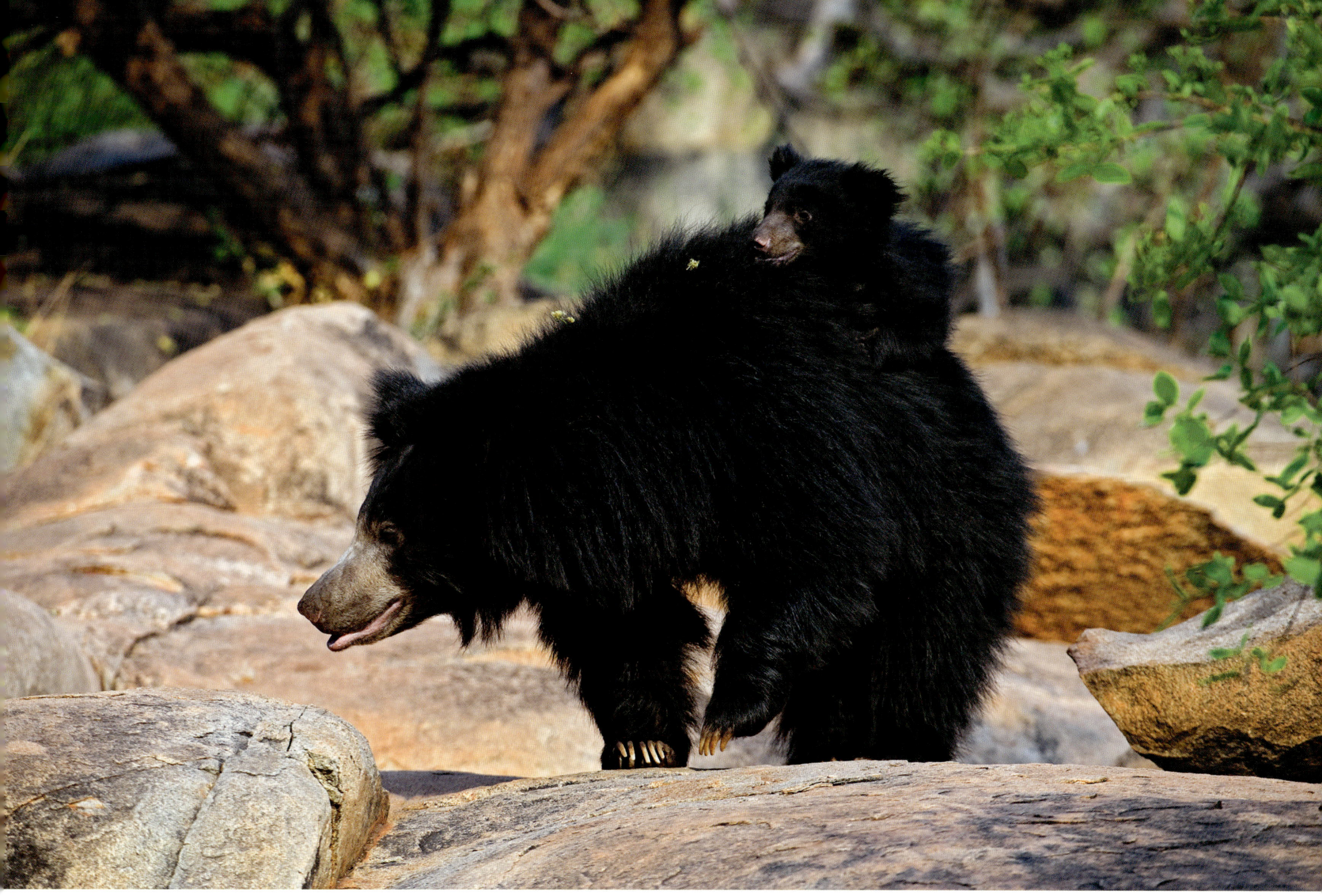

Sloth bears carry their young piggyback-fashion until they are about
nine months old. The cubs must learn to hold on to the mother's fur.

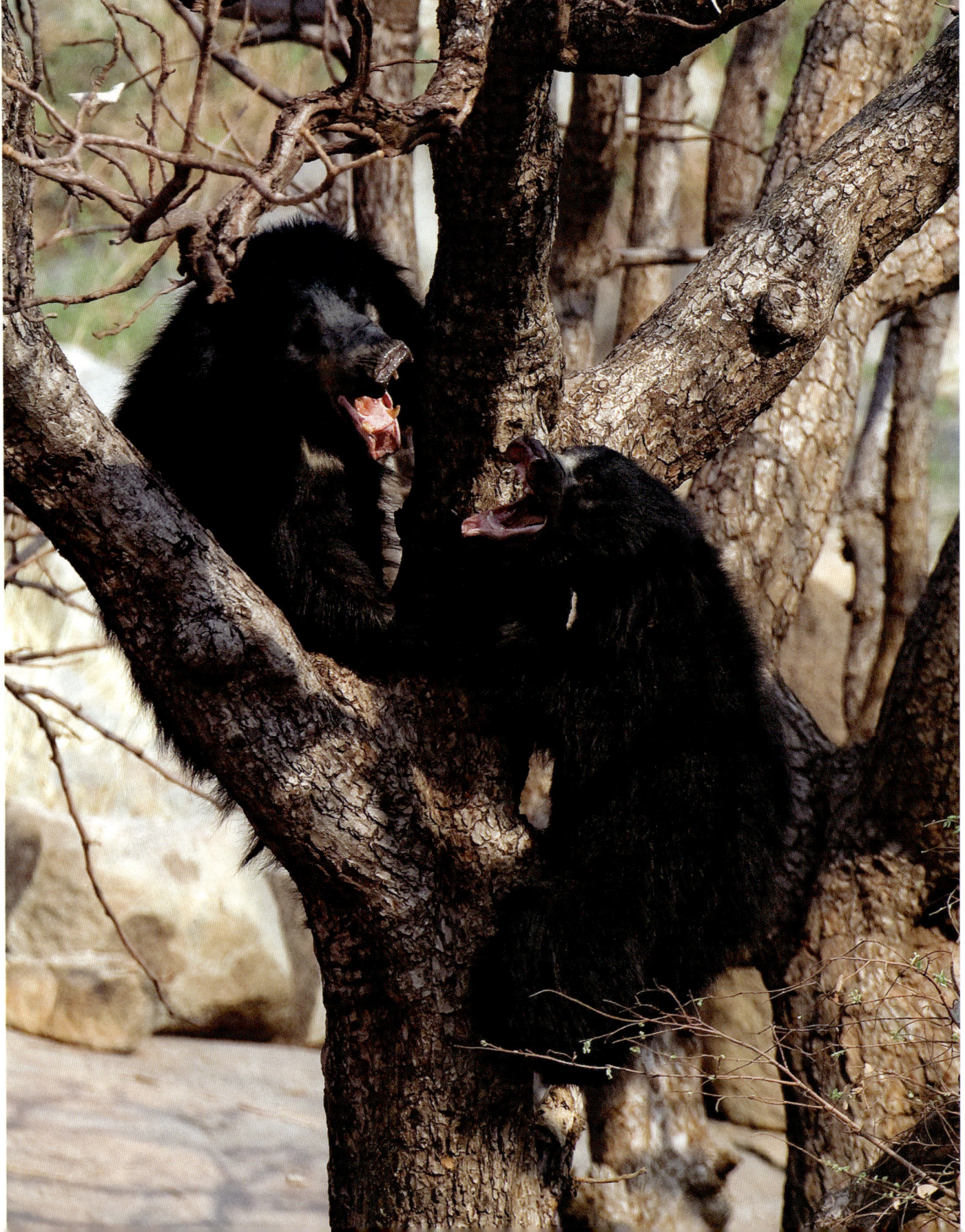

When young sloth bears start to explore the world around them, they do not venture too far from the protection of their mother (right). They have a lot to learn before they can fend for themselves. A young male climbs a tree and meets an old female (left). There is no room for either of them to step aside up there, so they both issue threats, using roars and bared teeth. That is enough to settle the argument and avoid injury.

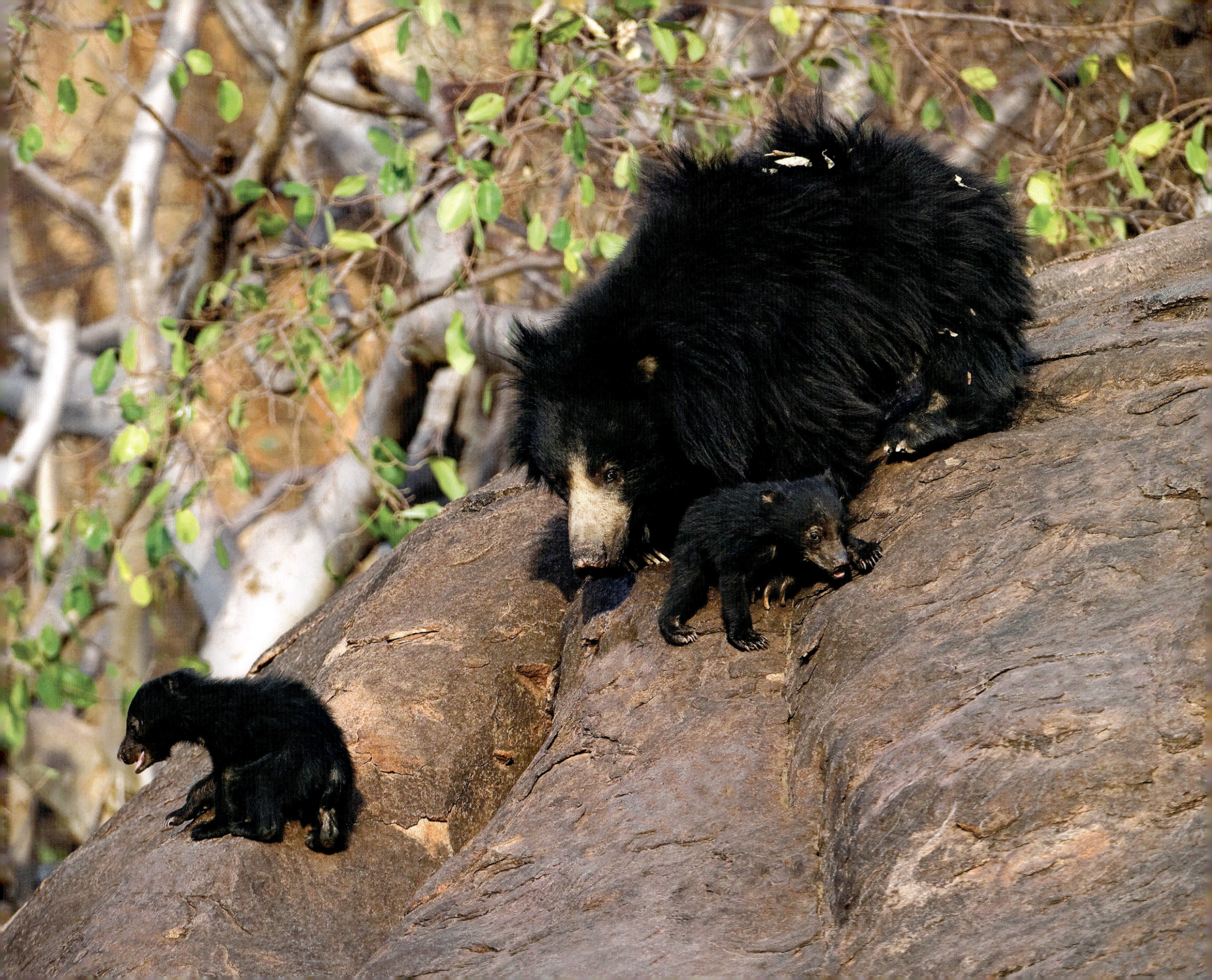

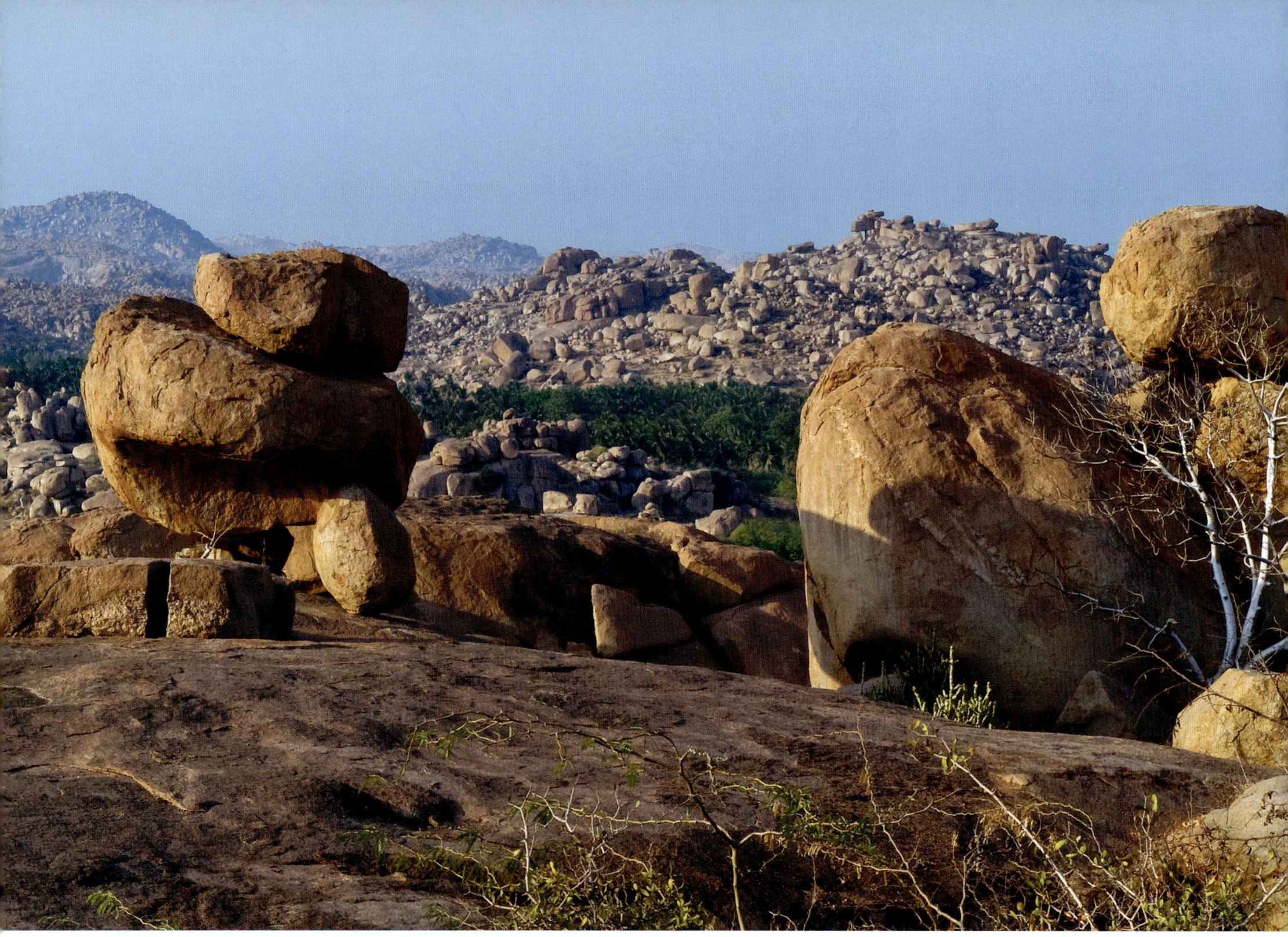

This bizarre rocky landscape in Karnataka is good sloth bear territory.
The many caves here offer shelter for the animals.

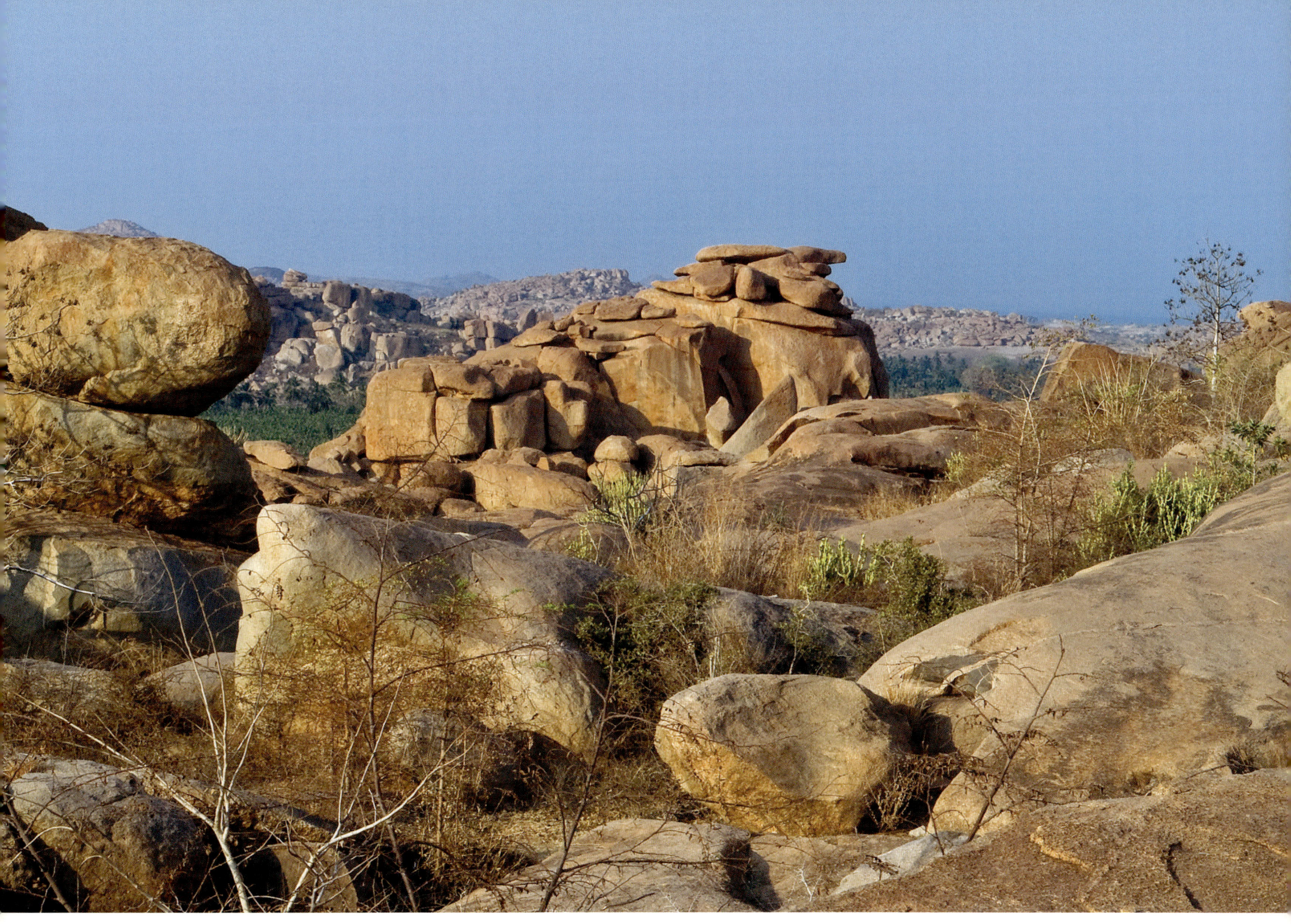

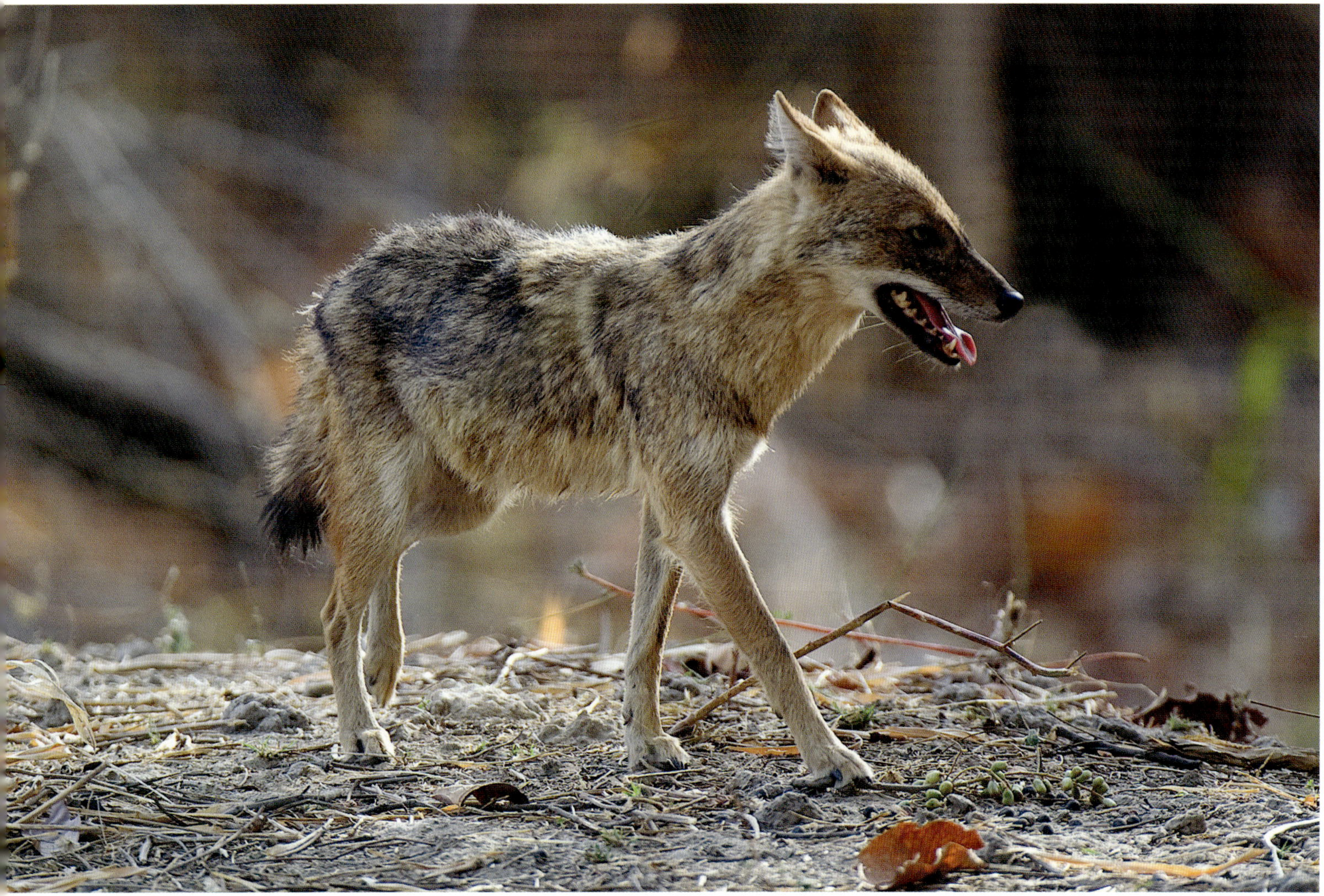

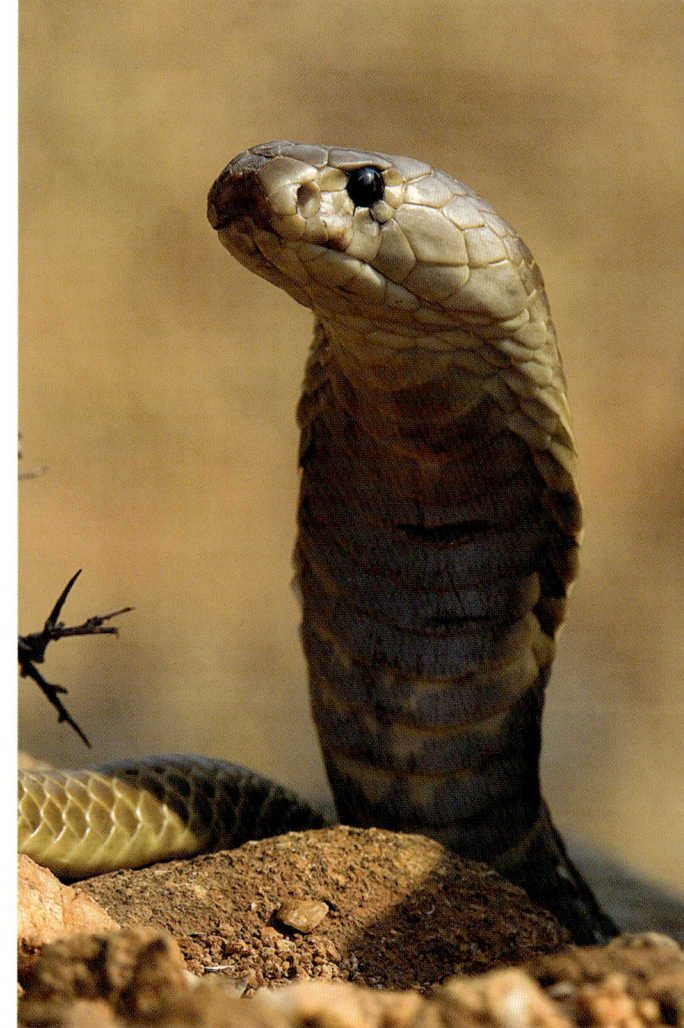

Many other animals share the bear's habitat. The ruddy mongoose (right) is especially keen on catching reptiles, even poisonous ones such as the Indian cobra (left). These snakes seek shelter in the many crevices among the rocks. The golden jackal hunts here as well (far left).

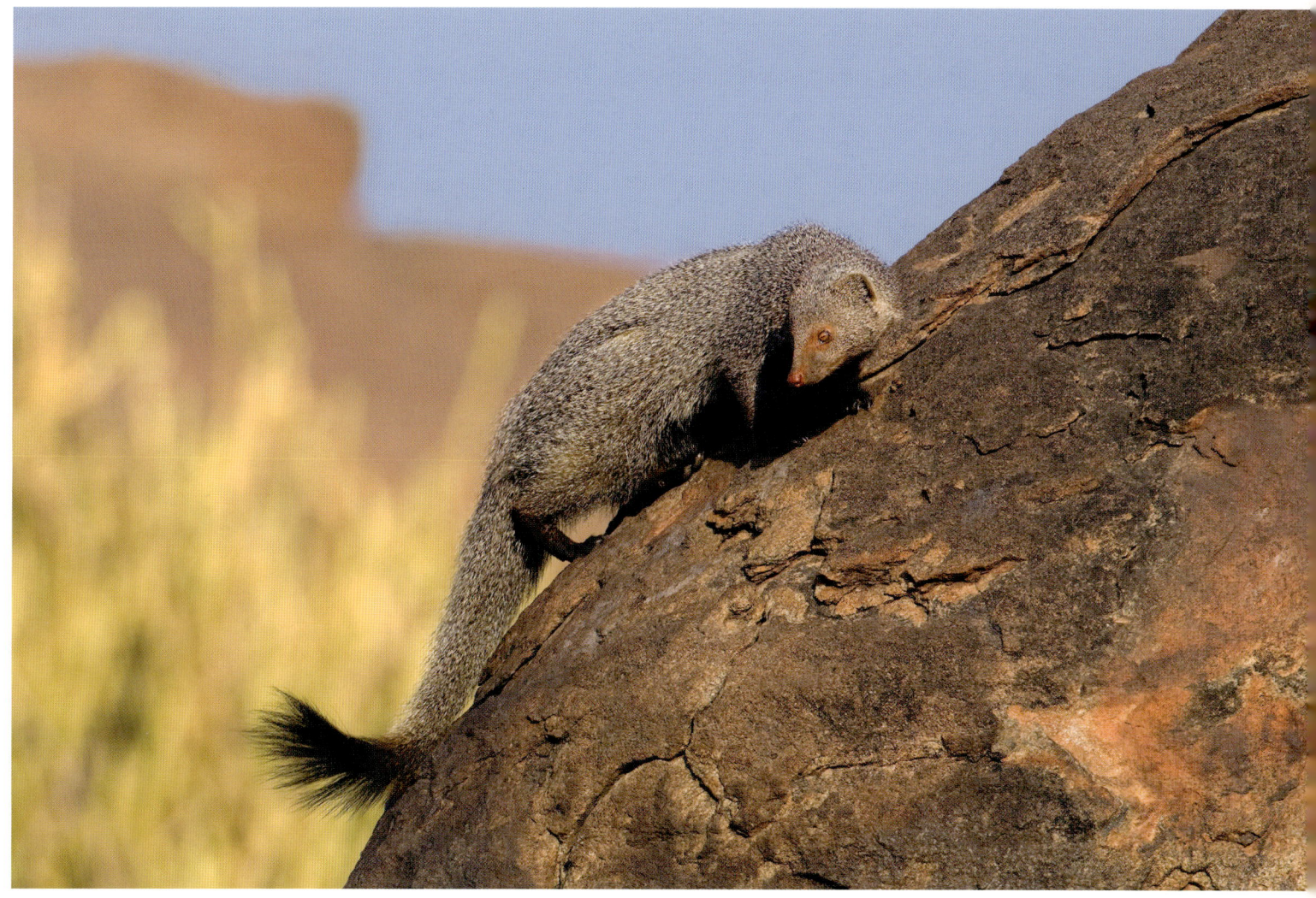

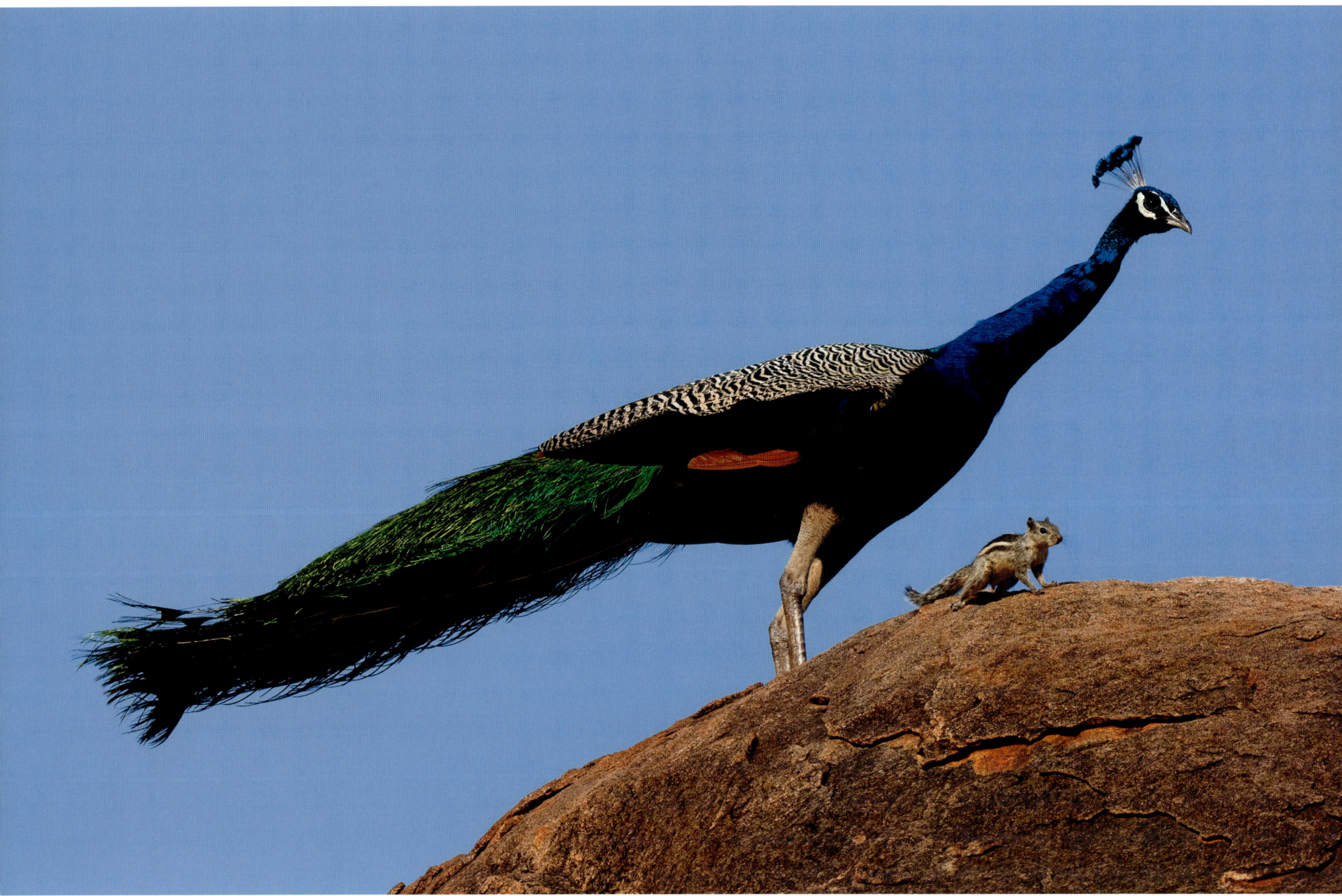

Peafowl and palm squirrels
inhabit the rocky terrain.
Among their neighbours
are bonnet macaques,
with their cheeky infants.
These monkeys live on fruit
and other plant material,
as well as the occasional
egg or insect.

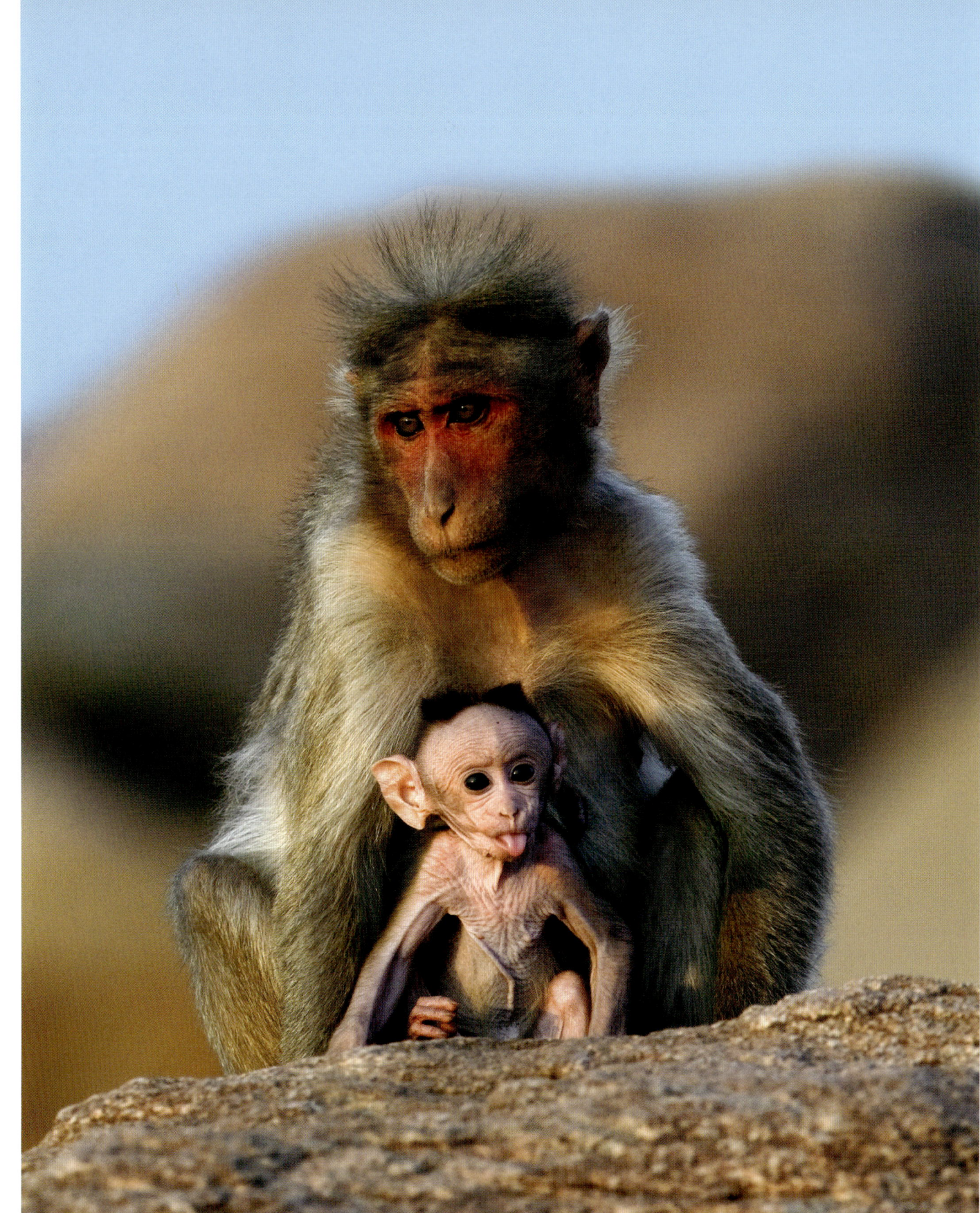

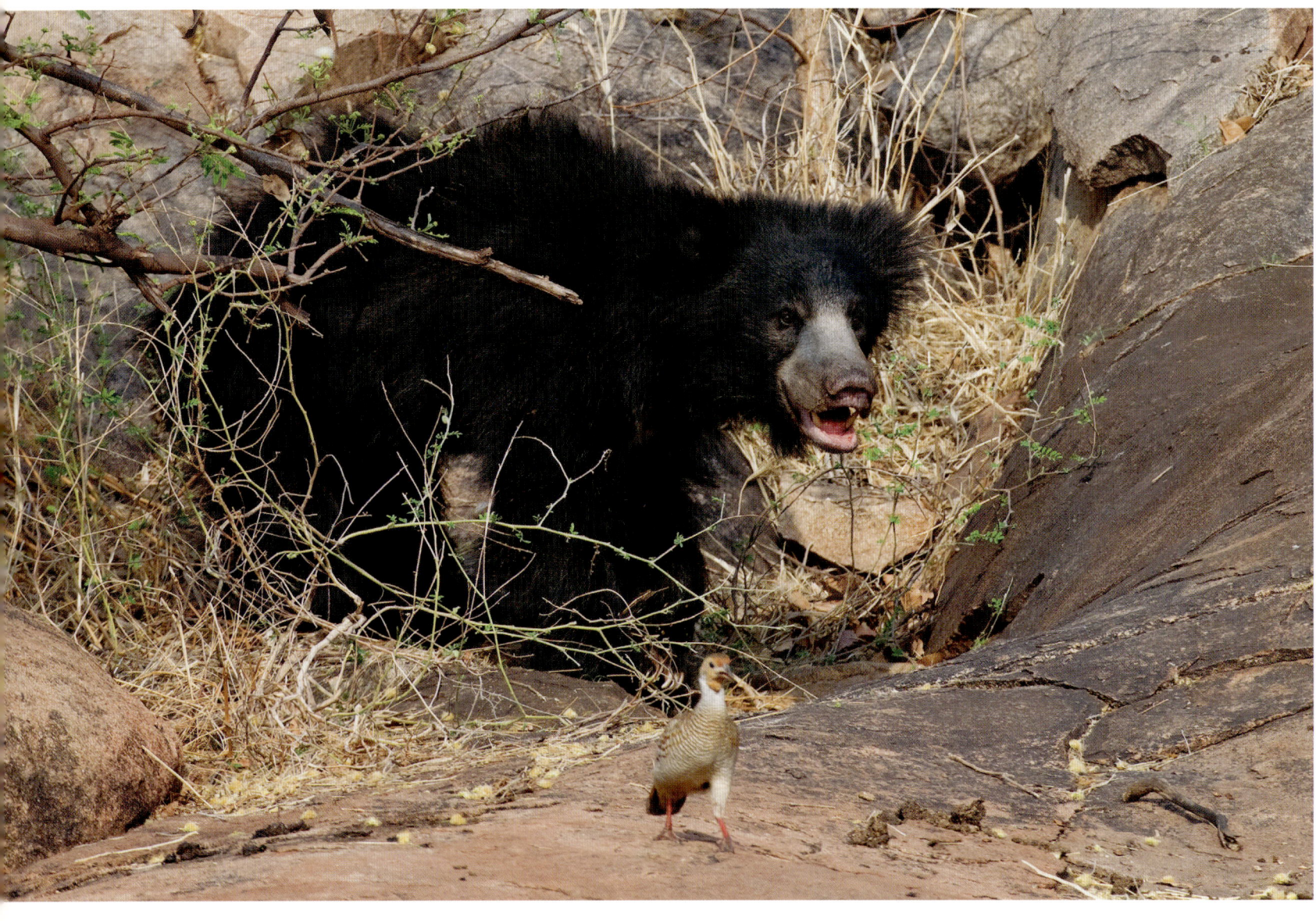

Two little bear siblings inquisitively explore the neighbourhood of the cave they were born in. They must learn the difference between friend and foe – but the peacock is completely harmless. Meanwhile mother keeps a wary eye on them, ready to step in should there be any danger. An Indian grey francolin (left) prefers to remain at a safe distance.

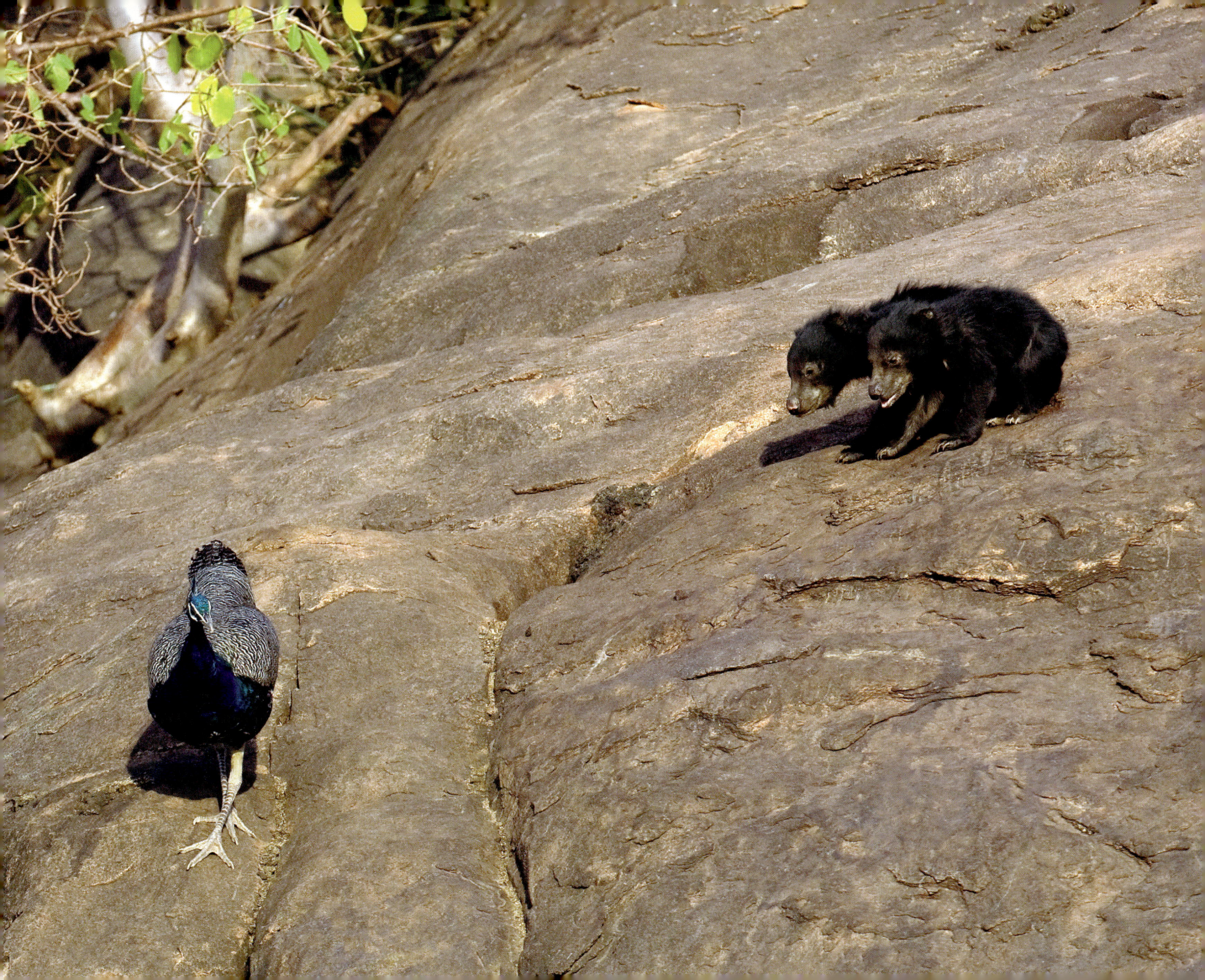

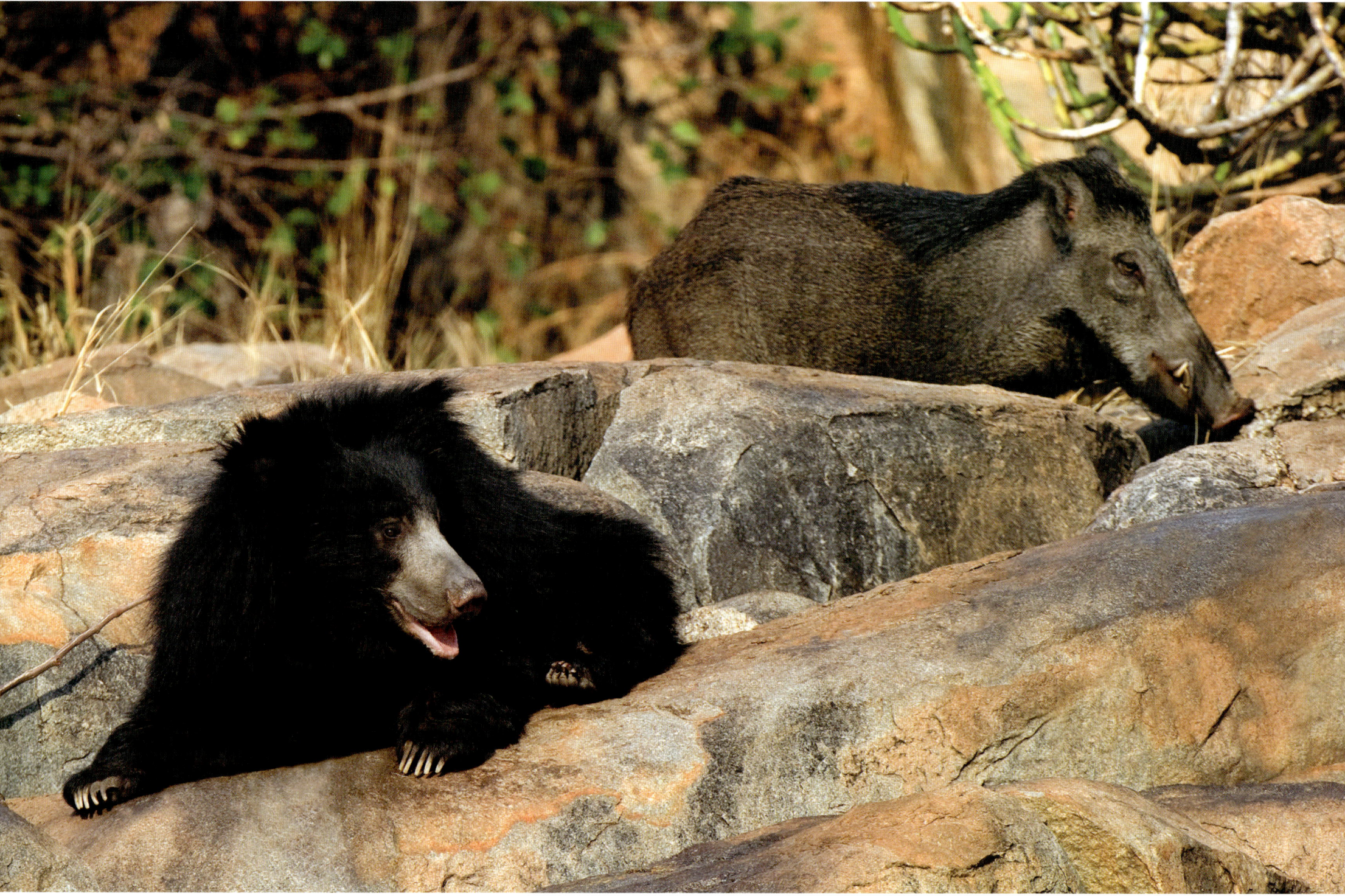

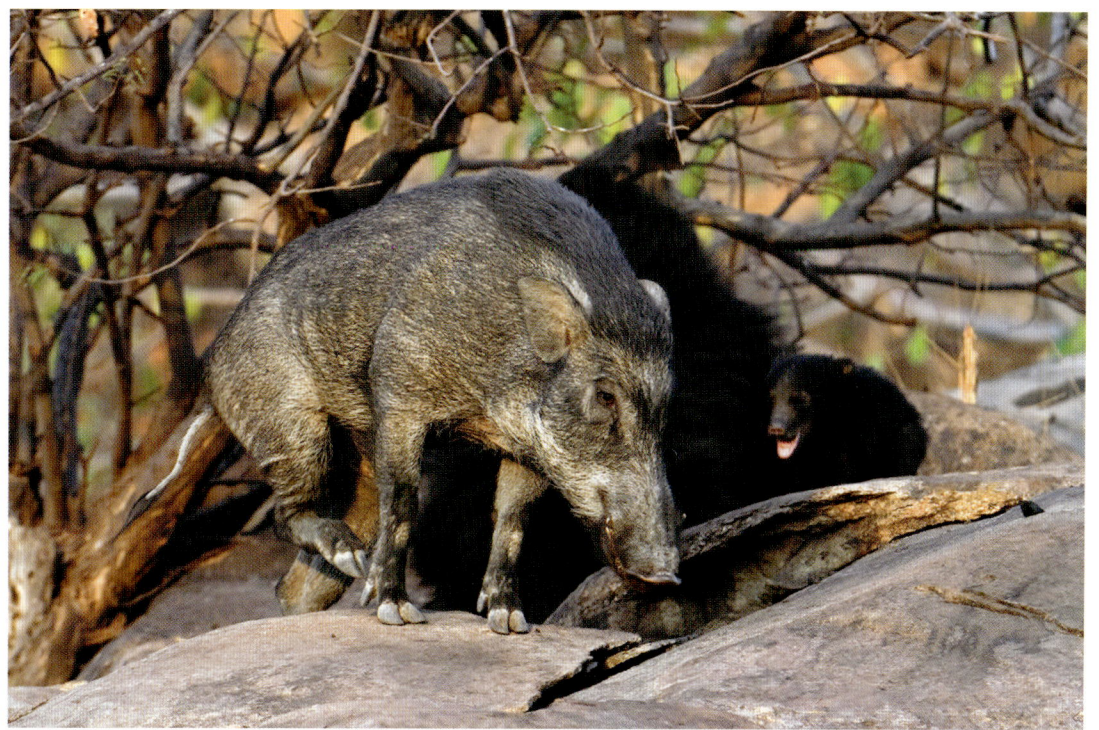

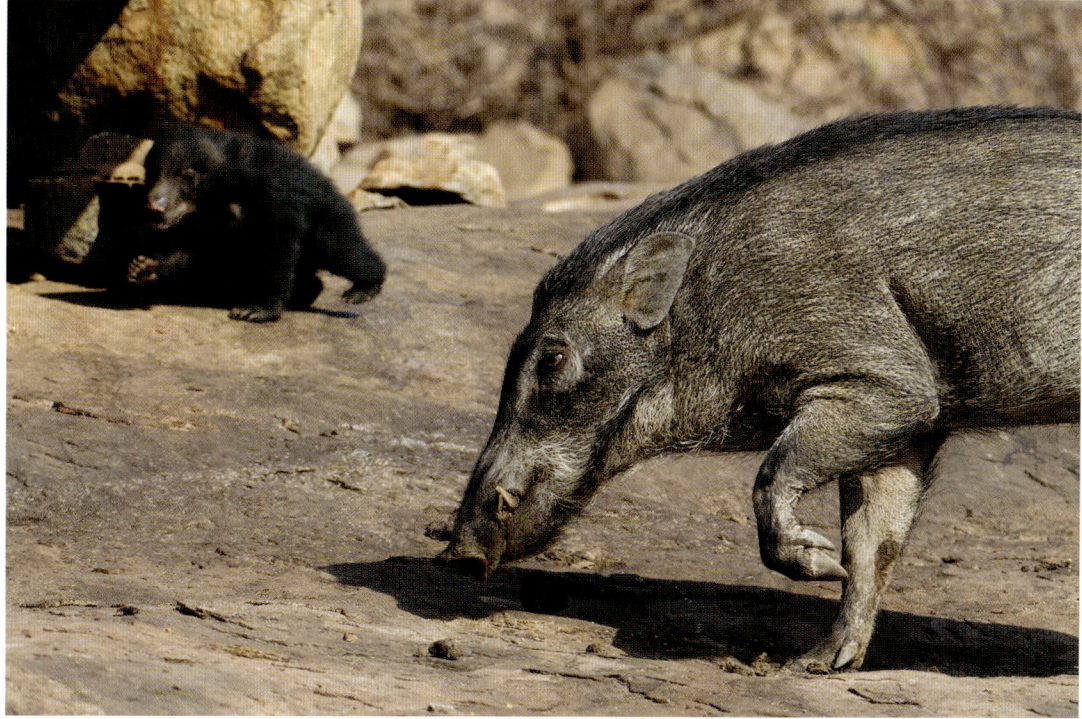

Sloth bears live side by side with wild boars. Adult animals tolerate each other, but these bear cubs have yet to learn that the self-confident wild boars are no danger to them.

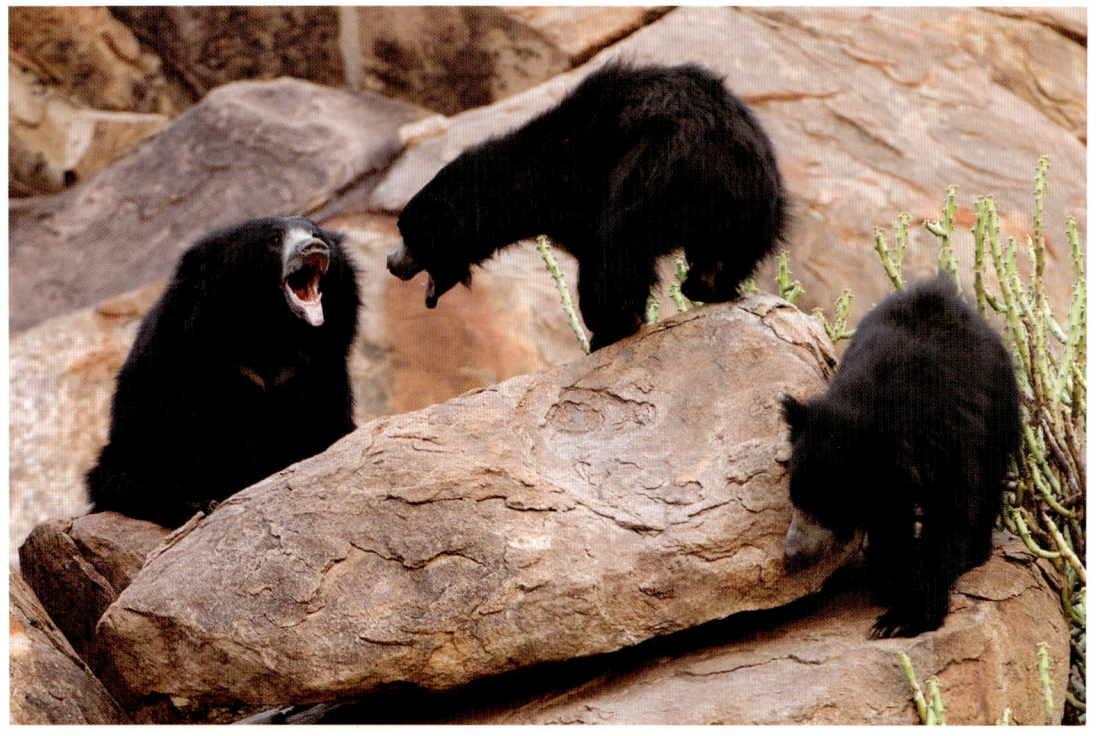

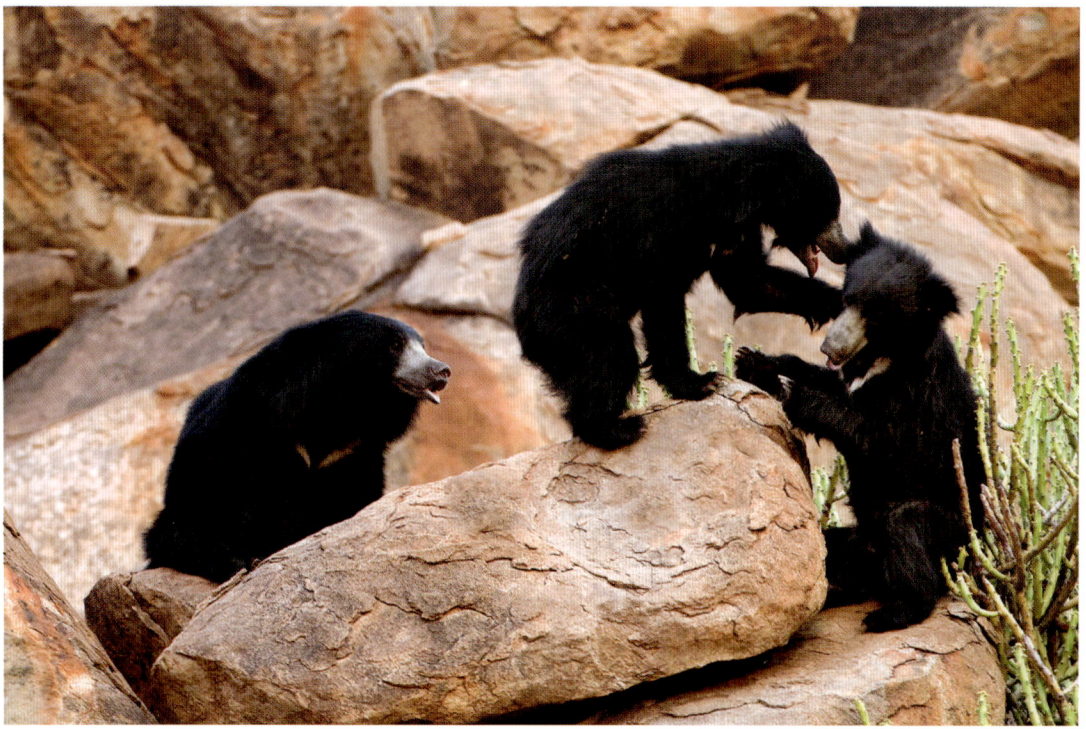

Little is known about the social behaviour of sloth bears, but two of my observations give interesting clues. In one case, a mother was seen defending her well-grown offspring, loudly and with great determination, against a strange male (right). The mother's reaction suggests that males do pose a threat to the cubs, although there are as yet no confirmed records of infanticide in sloth bears. However, this behaviour has been observed in species such as grizzly bears, tigers and lions, and most likely it occurs in sloth bears as well.

On another occasion, an encounter between an old male and two younger males appeared quite aggressive (top left). But then the old bear observed the younger ones as they played – he was most likely their father (bottom left). This peaceful encounter suggests that in sloth bears long-term family ties exist even after the young have become independent. The behaviour seen here was previously unknown, and it is indeed surprising, since the males seem to play no part in raising the cubs.

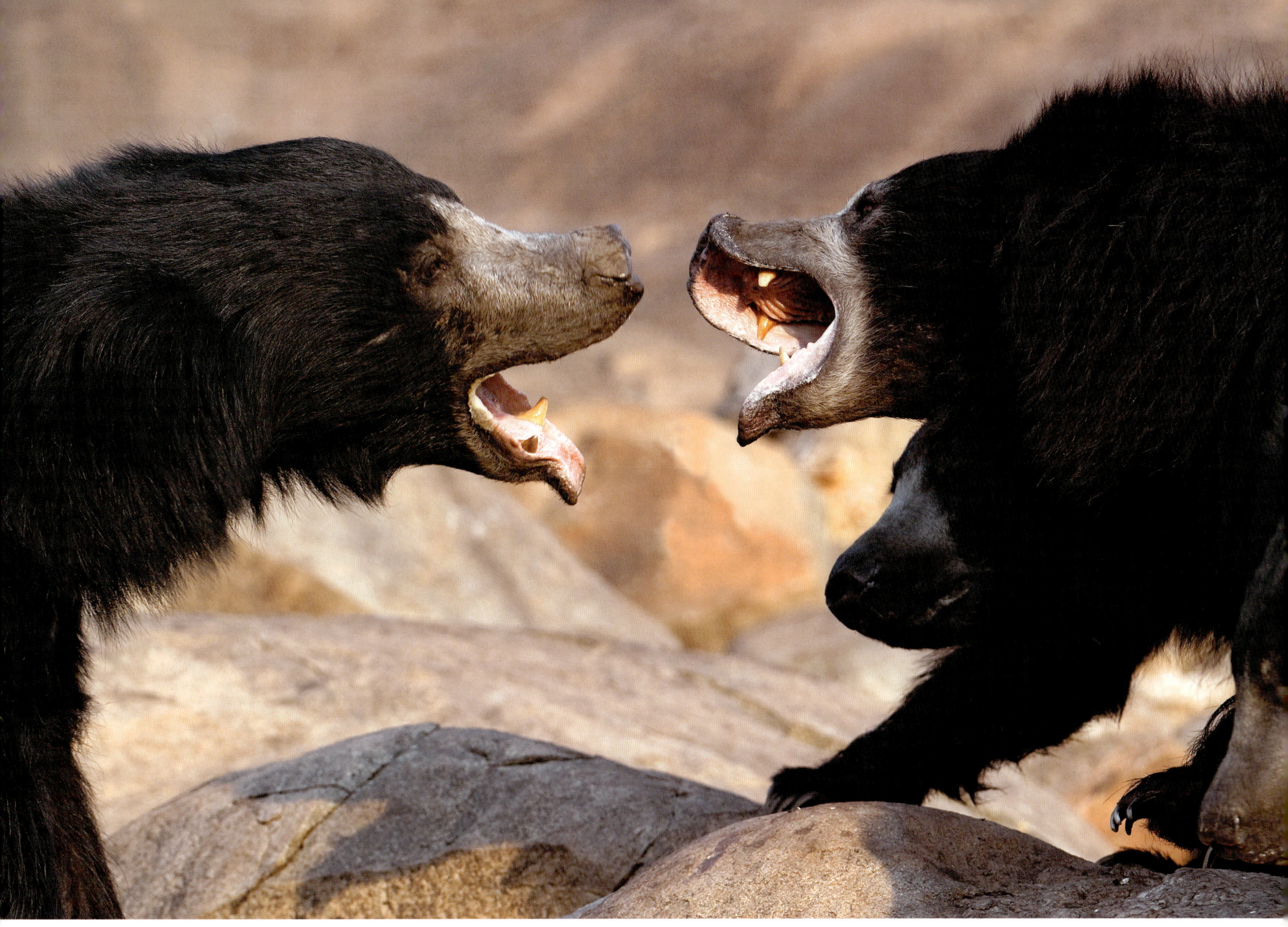

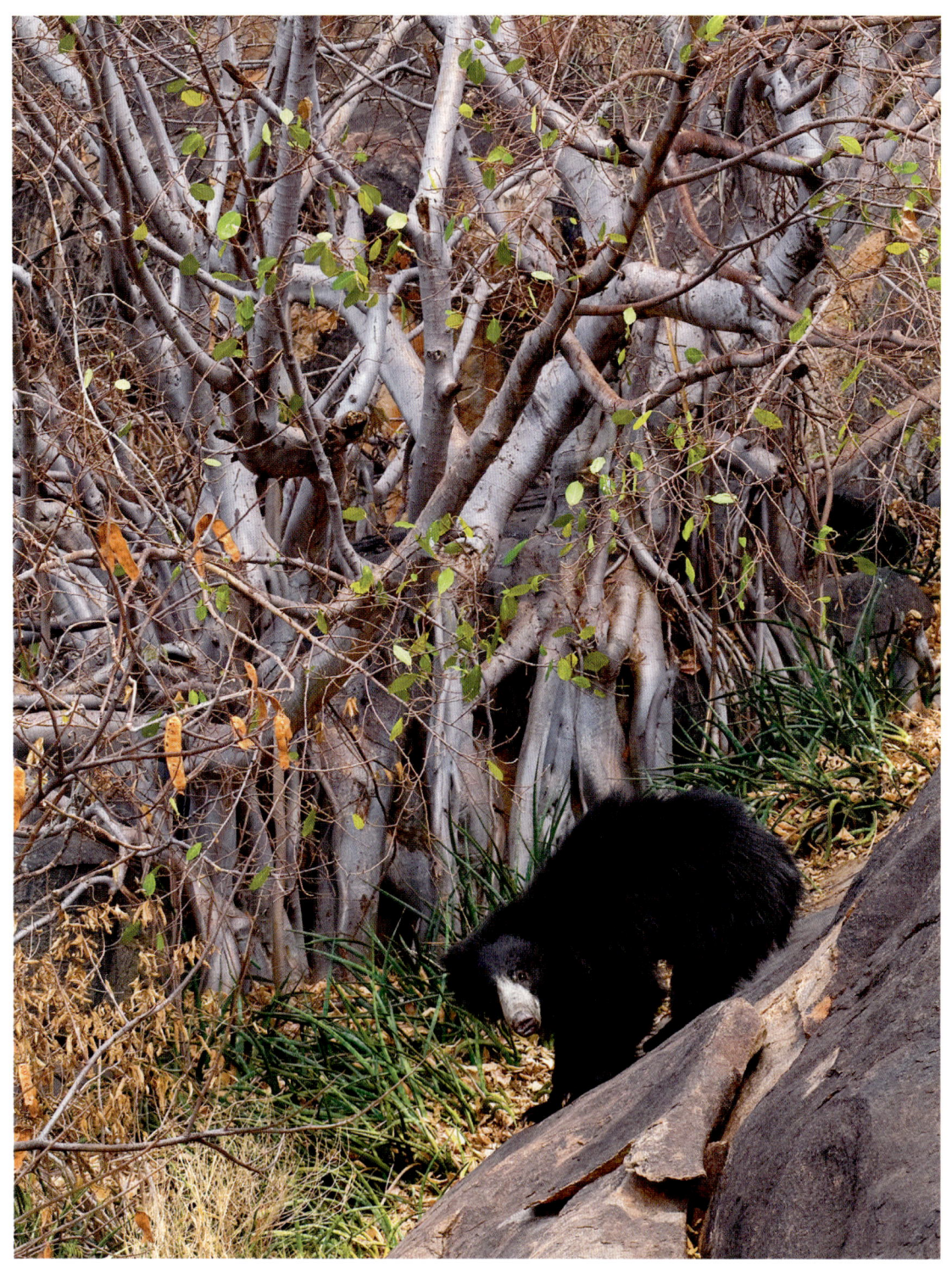

Sloth bears are endemic to the Indian subcontinent. They live in a variety of habitats, including grassland, thorn scrub and jungle. Historically, their range extended from the southern foothills of the Himalayas to Sri Lanka. Due to habitat loss, human encroachment and poaching they have disappeared from a large part of their former range, and today their distribution is patchy. However, the rocky hills of Karnataka still support a healthy population of sloth bears.

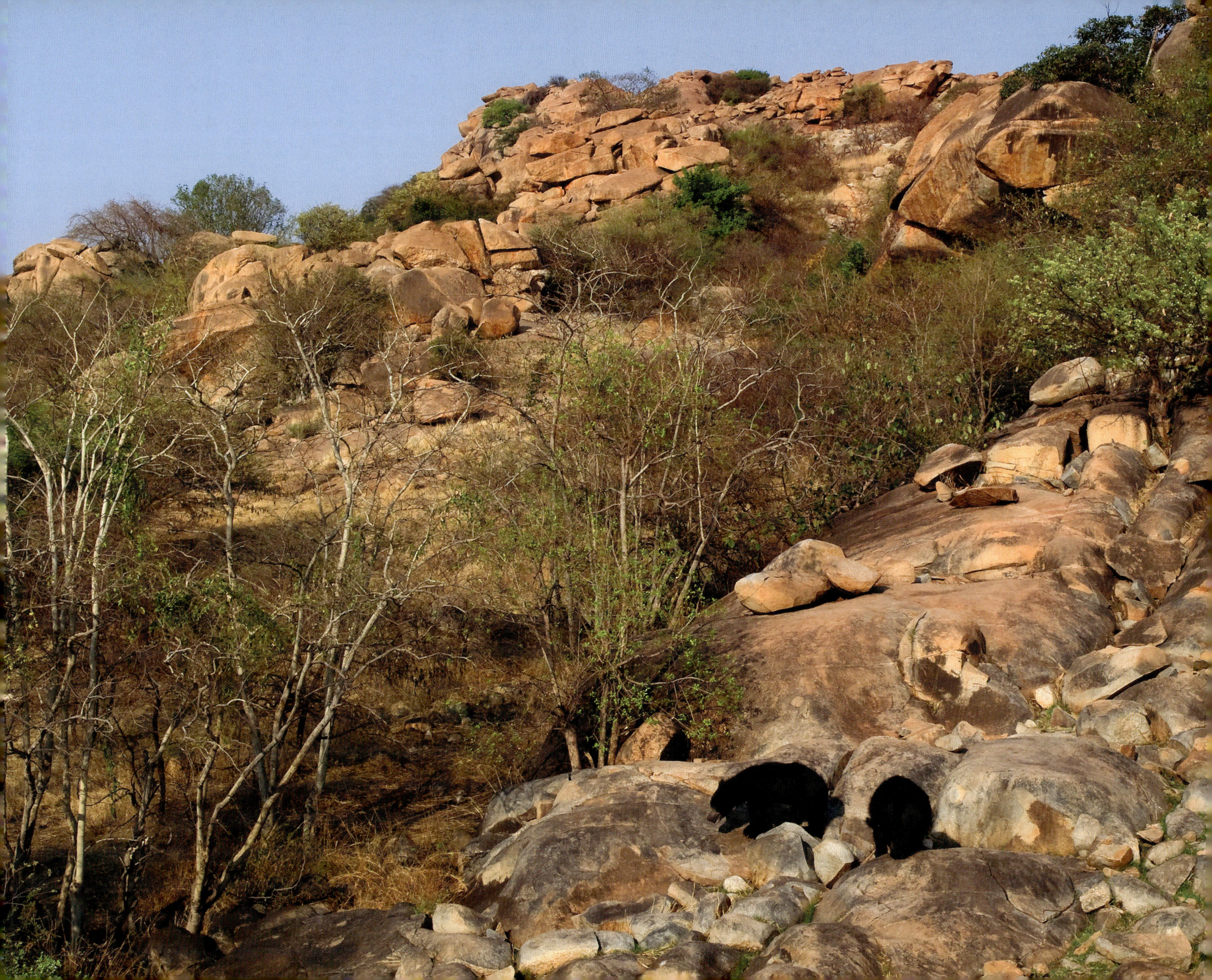

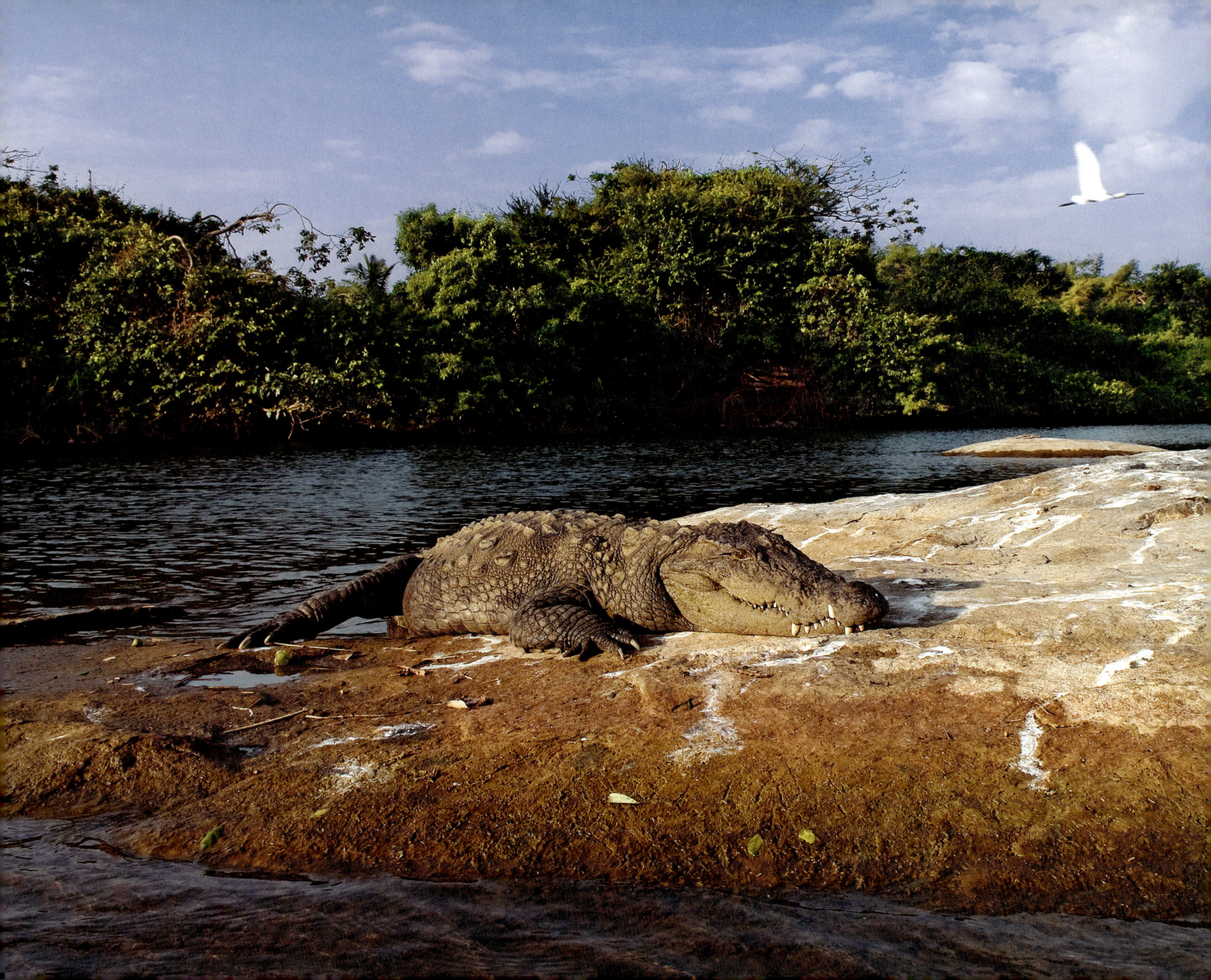

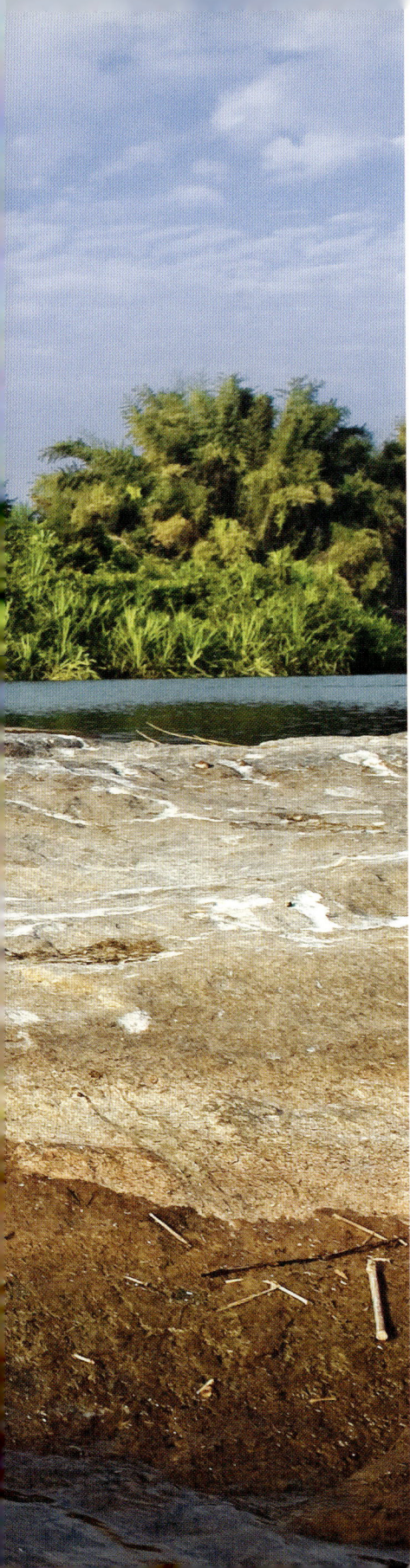

In the Swamps

India

For months the hot, tropical sun beats down on India, and large swaths of the country become as dry as dust. But in the wetlands, still awash with water, there is burgeoning life.

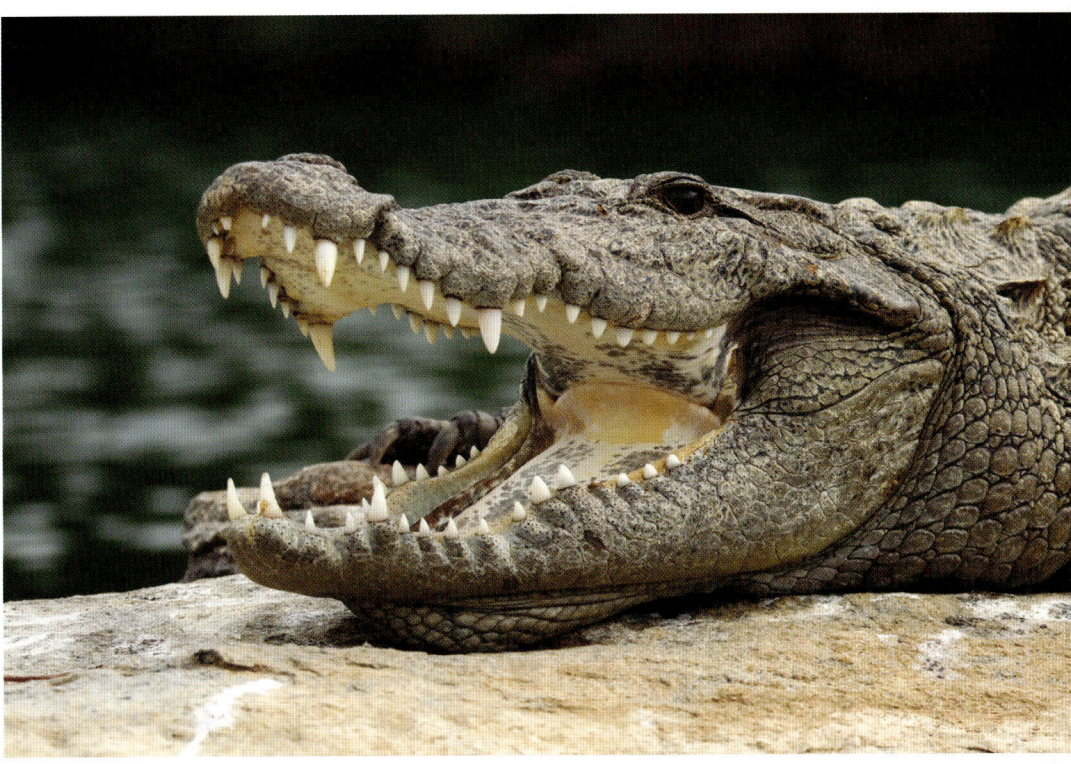

Water is Life

Our boat glides slowly across the Kaveri River in Karnataka. The boatman knows that any sudden movement would drive away the crocodiles, some of which are basking on small, stony islands in the river. Birds stream in flight above our heads or perch on bushes on the riverbank, among them a variety of herons, storks, spoonbills, ibises and pelicans. Countless calls mingle in riotous concert. Fruit bats hang in the upper storeys of the trees. Although they normally sleep through the daylight hours, there are always a few that squabble with their neighbours over the best roosting spots. When an animal is unlucky enough to fall into the water, it is quickly devoured by a crocodile. To all appearances those reptiles slumber on their rocks, but in reality they are alert, waiting patiently for an opportunity. In the meantime we have approached to within a few metres of a marsh crocodile. The dagger-like teeth that protrude from its jaws are a constant reminder to keep a safe distance.

For a long time crocodiles were the uncontested rulers of tropical swamps. The giant reptiles have scarcely changed in the 230 million years of their existence. Already an ancient lineage during the heyday of the dinosaurs, they are first-class products of evolution. In the modern world, however, they have ever fewer refuges. Once, the marsh crocodile could be found in all Indian river systems and in many freshwater lakes, but their populations are now drastically reduced. These animals can tolerate famine and drought for astonishing lengths of time, but they cannot cope with the permanent transformation of their habitats.

The birds have an easier time of it. If their habitat is destroyed, at least they can take to the air and search for a new home. Normally the dry season ends with the onset of the monsoons, the rain-bearing winds. In recent years, however, they have brought much less precipitation than hoped for in some regions – such as Keoladeo National Park. It was the bird diversity of this conservation area, more widely known by the name of the neighbouring city of Bharatpur, that led to its establishment as a UNESCO World Heritage Site. But a series of dry years quickly transformed the area: water bodies shrank to a fraction of their former size, birds stayed away or left, and cows and goats consumed a large proportion of the vegetation. Bharatpur was hardly recognizable, and seemed to have lost all entitlement to its designation as a World Heritage Site. Although it was possible to divert water from a reservoir into the national park, the farmers resisted, for they too suffered from the drought. Without water, their fields produced no crops. Whose rights are more important in times of need? Those of the people or those of the wild animals? In the end, while farmers, authorities and conservationists argued, a violent monsoon solved the problem. Heavy rainfall provided water for all. The farmers could harvest their crops again, and the bird paradise recovered. Since then, however, water has had to be pumped into the national park on many occasions.

The example is a reminder of how fundamental water is for all life on earth. In a time of global climate change, it is more difficult than ever to calculate whether this resource will continue to be available in sufficient quantities. In just a few short years, water stress can transform flourishing landscapes, be they inhabited by people or by animals, into wastelands.

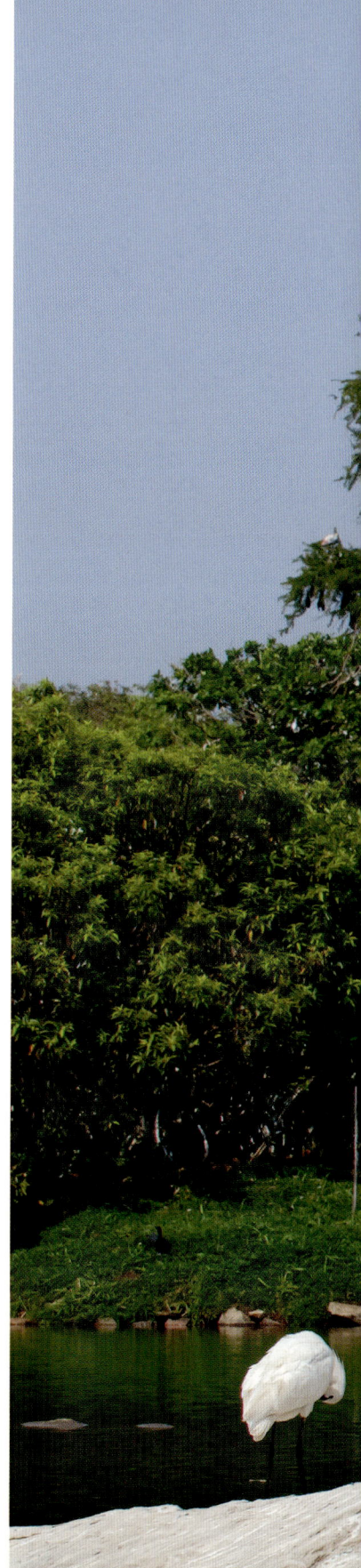

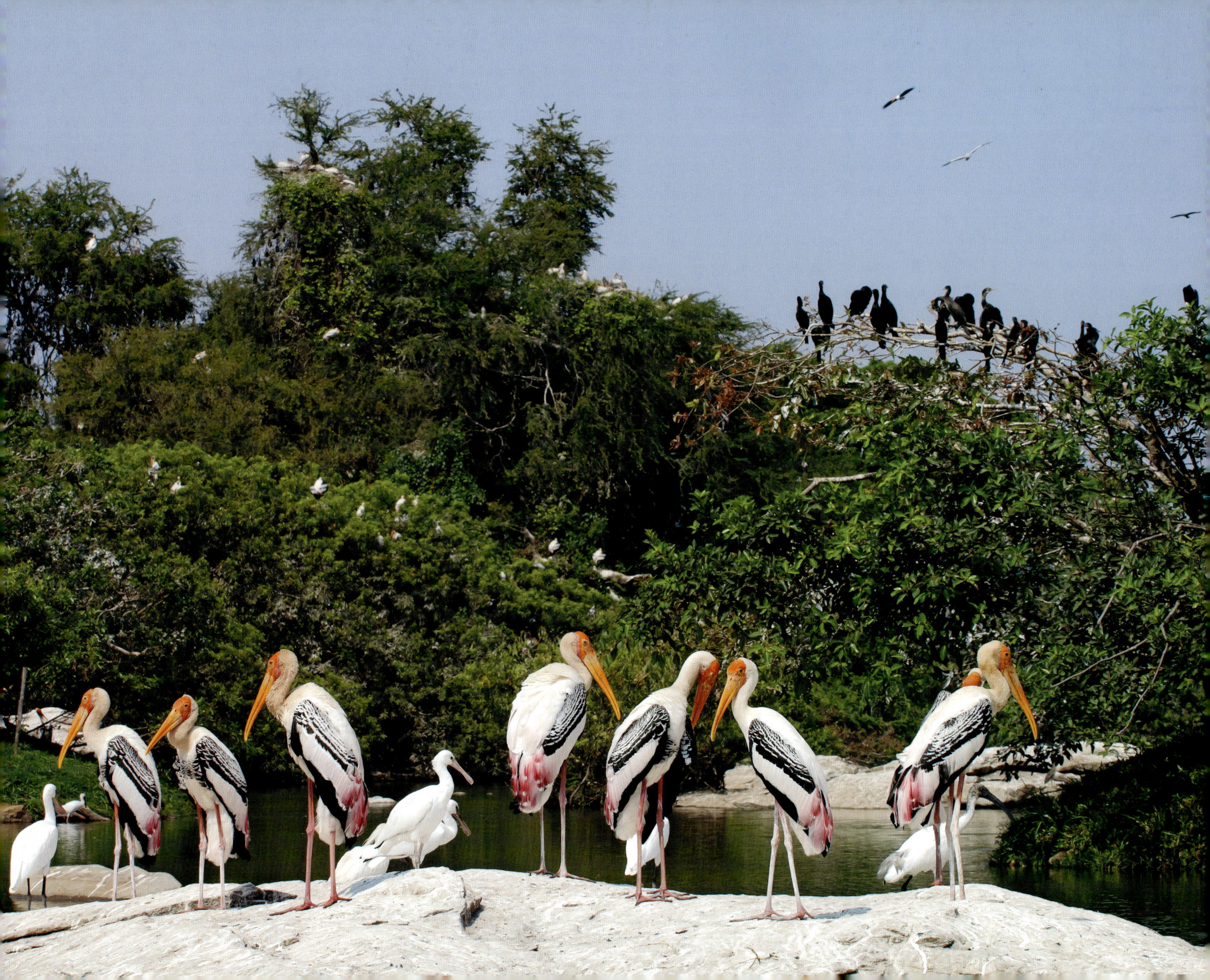

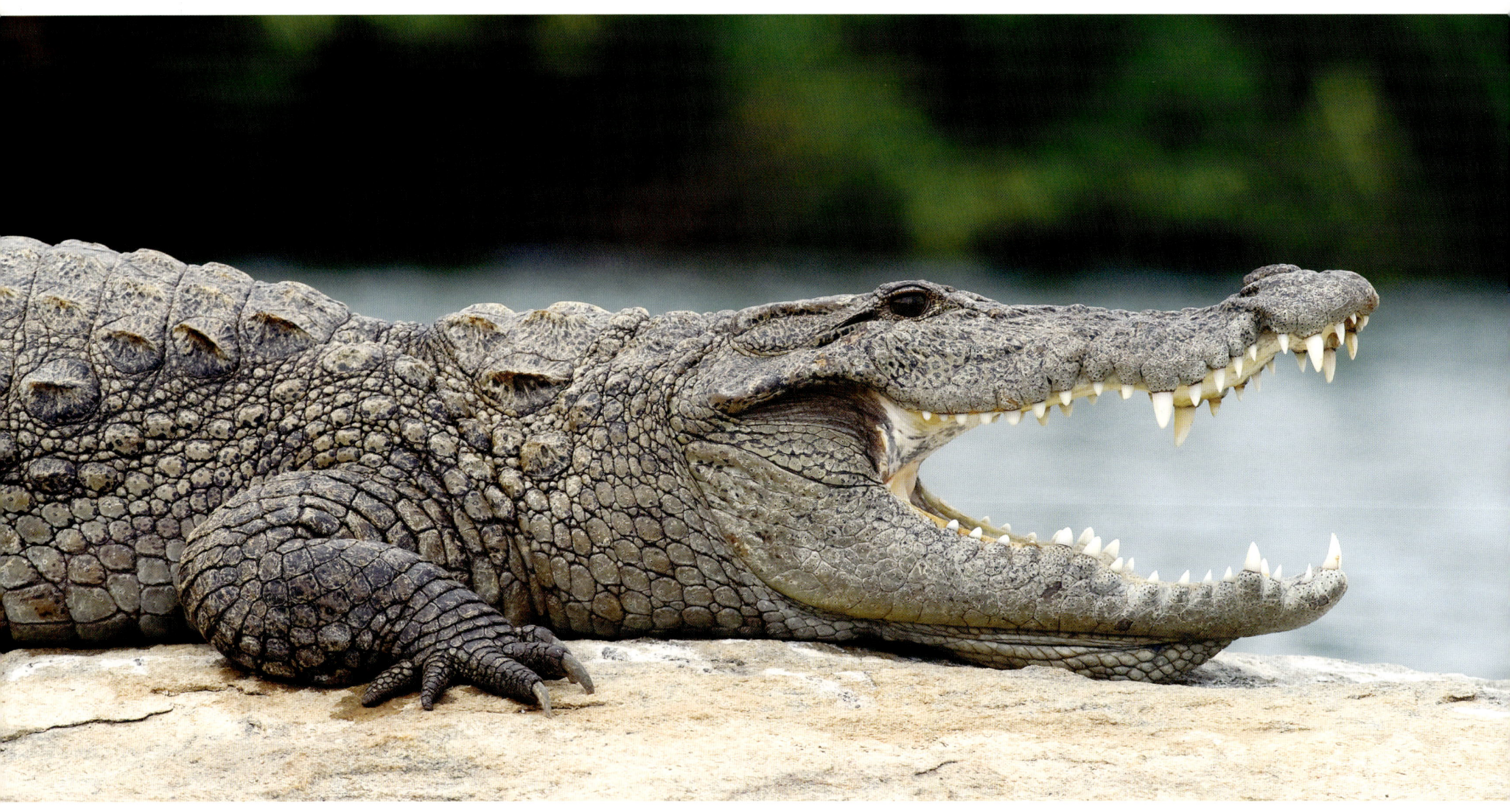

From time immemorial crocodiles have ruled over tropical rivers and lakes. In India, the marsh crocodile or mugger (above) is the most widespread species. A large specimen can reach a length of four metres (13 feet). Another reptile that likes to hunt in swampy areas is the water monitor (right). With a body length of about two metres (6½ feet), it is also an impressive sight. In India, the water monitor occurs only in the eastern parts of the country.

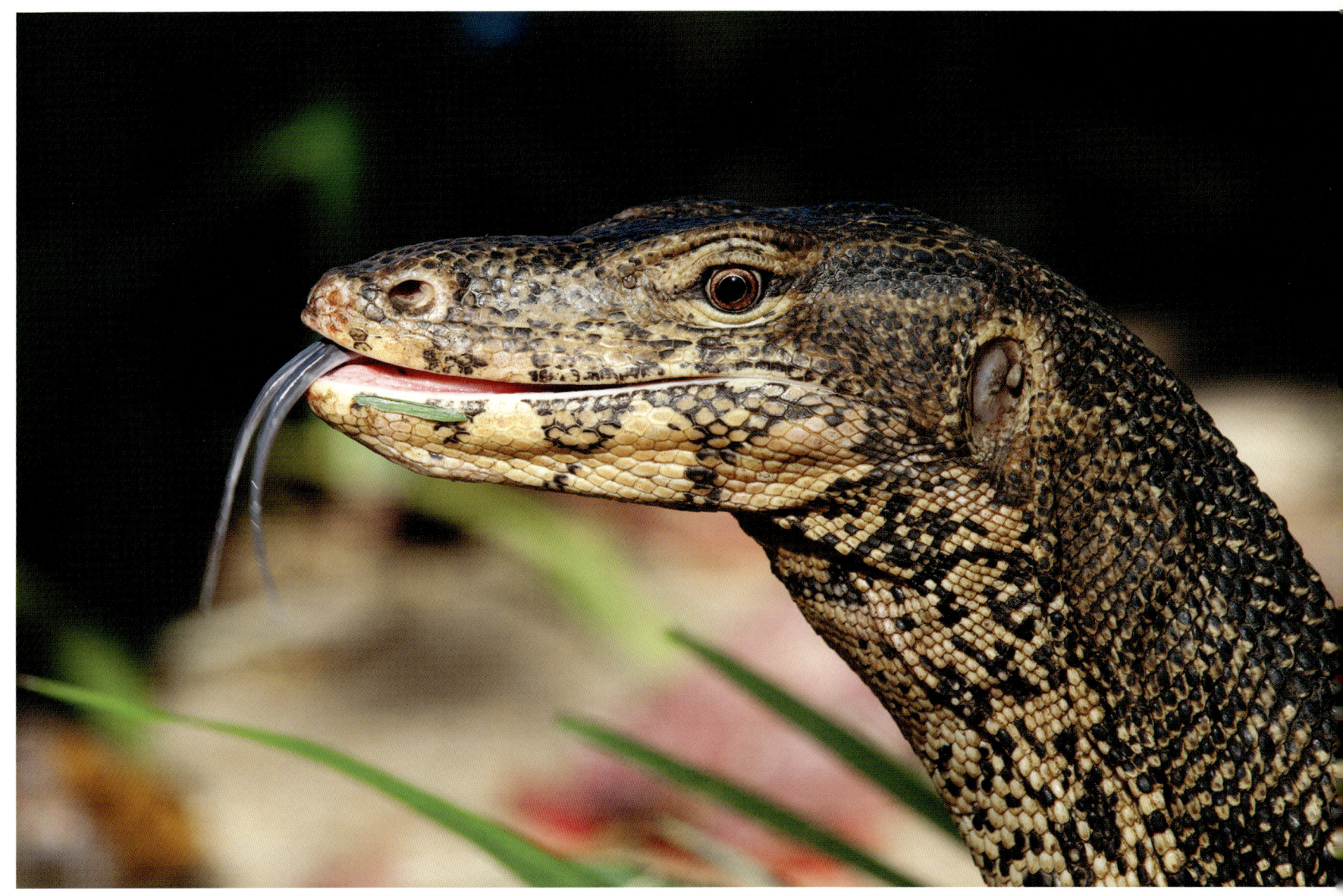

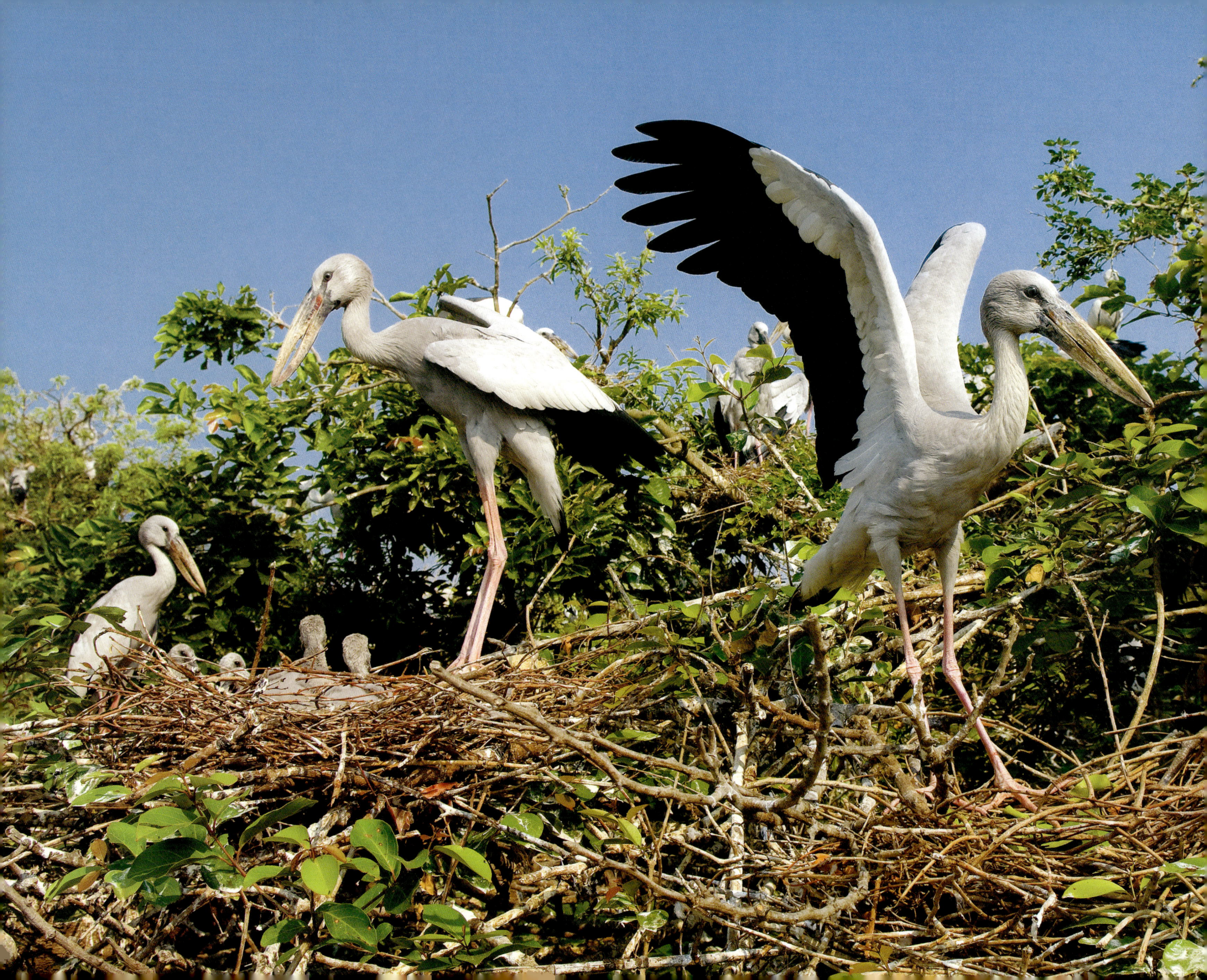

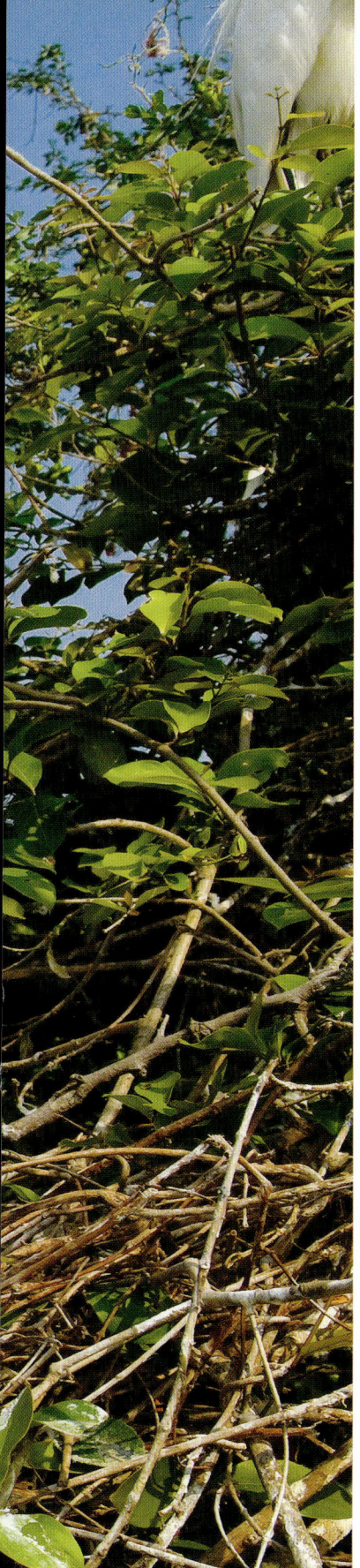

Asian openbill storks breed
in large colonies, for
instance on the Kaveri River
in Karnataka (left). Cattle
egrets (right) live in the
same area. The plumage of
this usually snow-white bird
turns partly orange in the
breeding season, and even
the bird's eyes and beak are
now strikingly colourful.

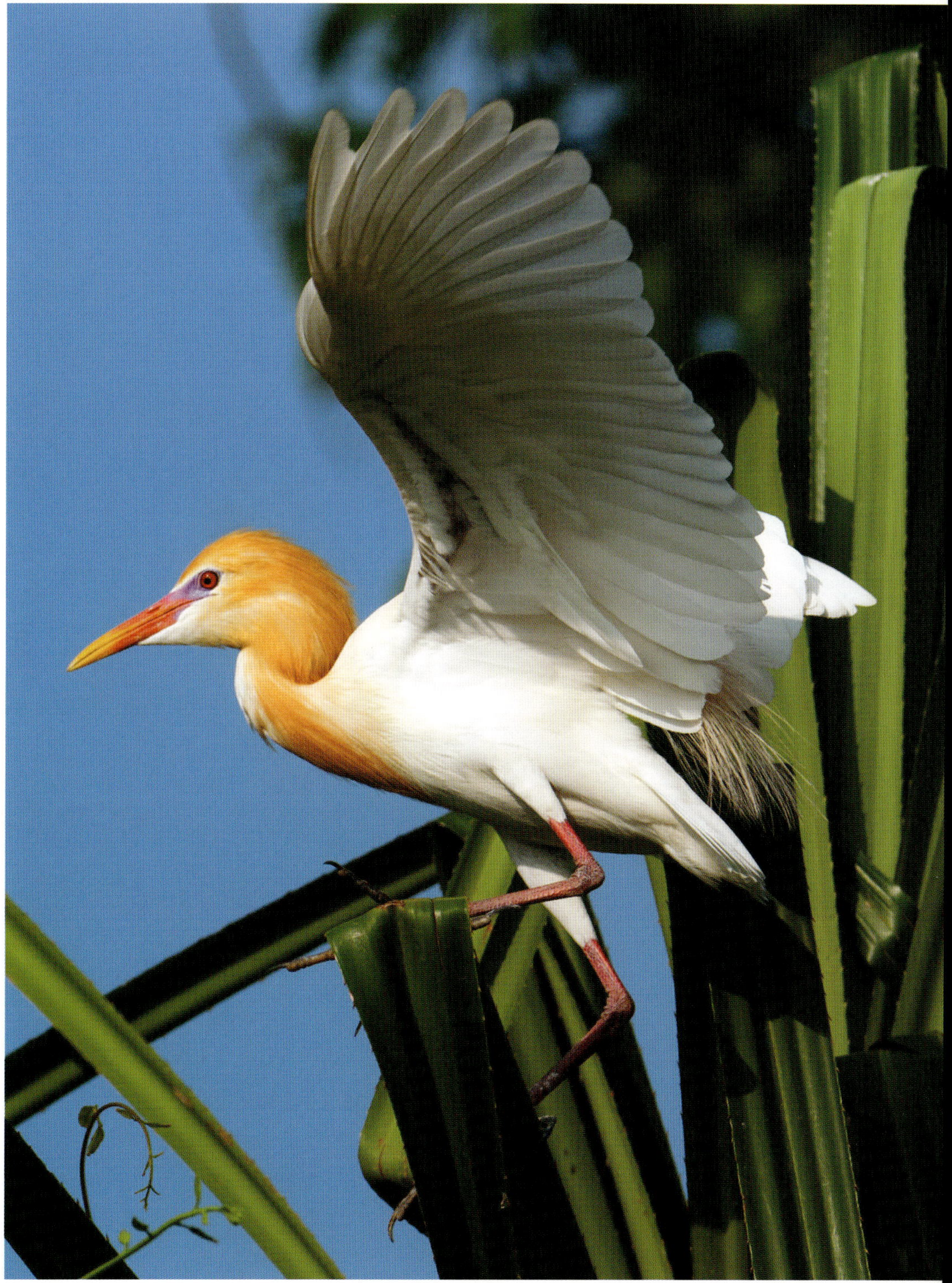

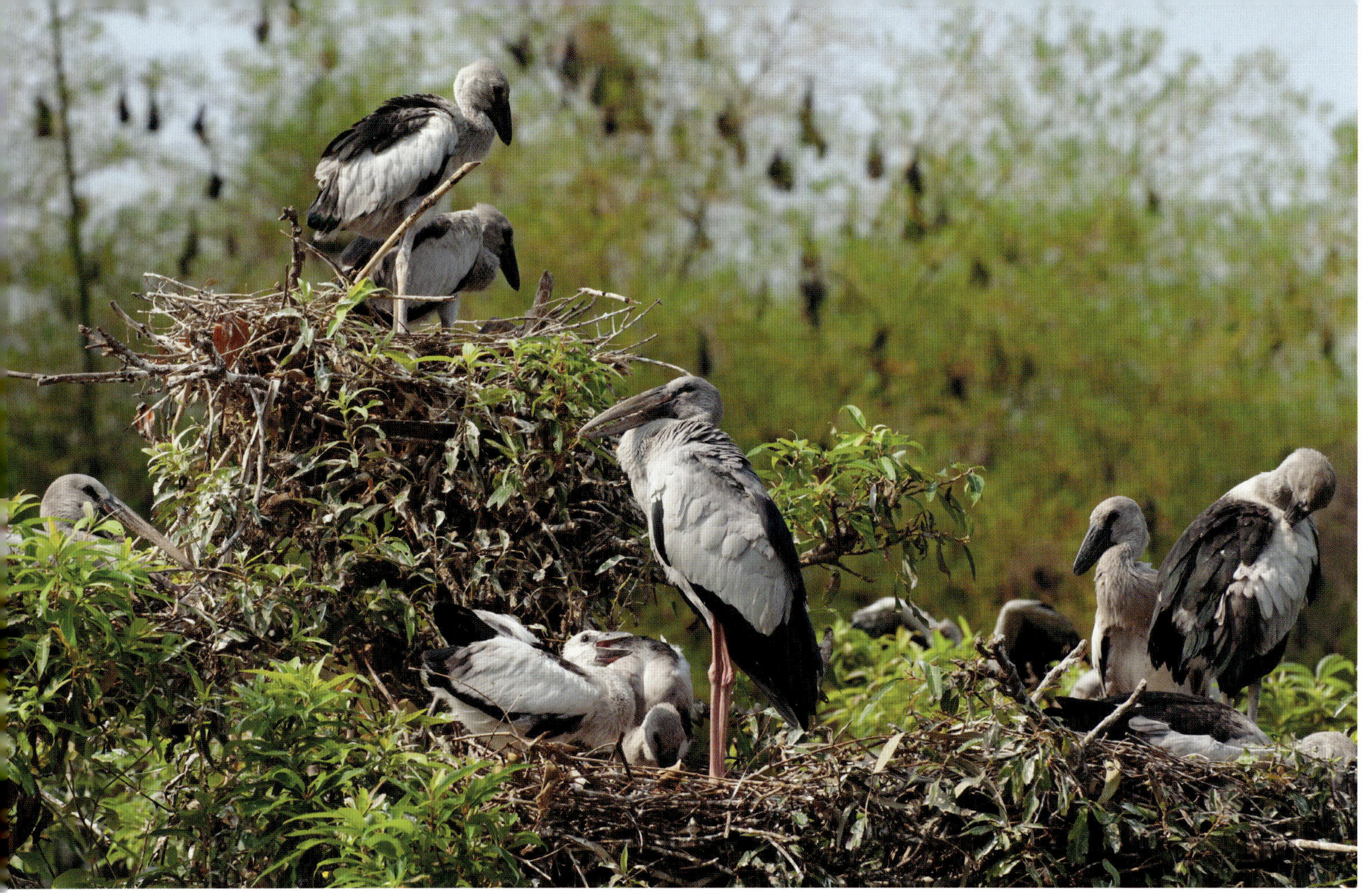

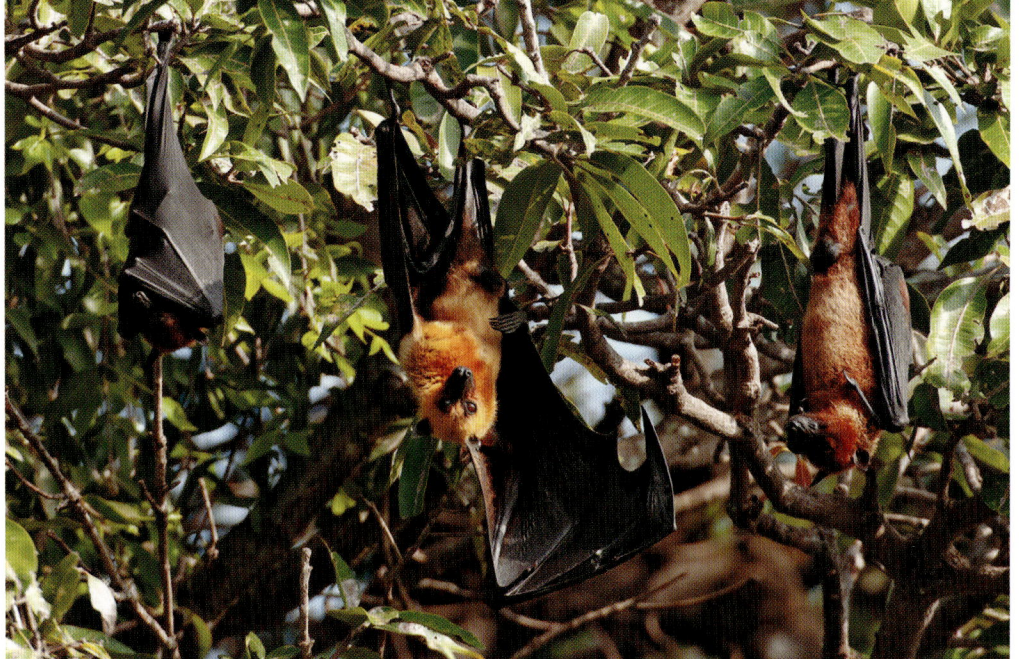

Behind a breeding colony of Asian openbill storks, there is a roost of Indian flying foxes. These large bats are hanging upside down in the vegetation, where they sleep through the day. In the evening they swarm out from their roost in search of fruit-bearing trees. The wingspan of an Indian flying fox can reach 1.2 metres (4 feet).

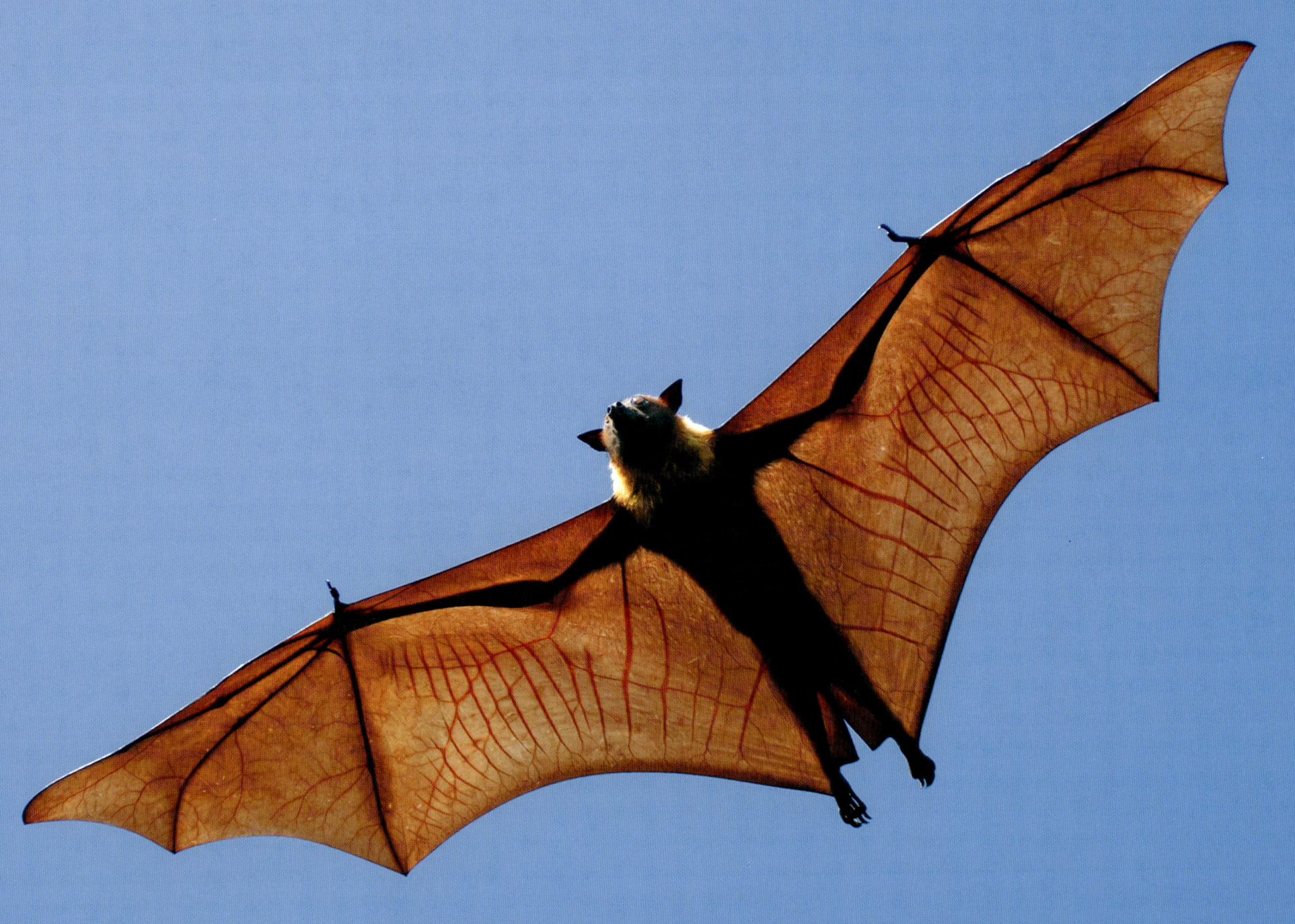

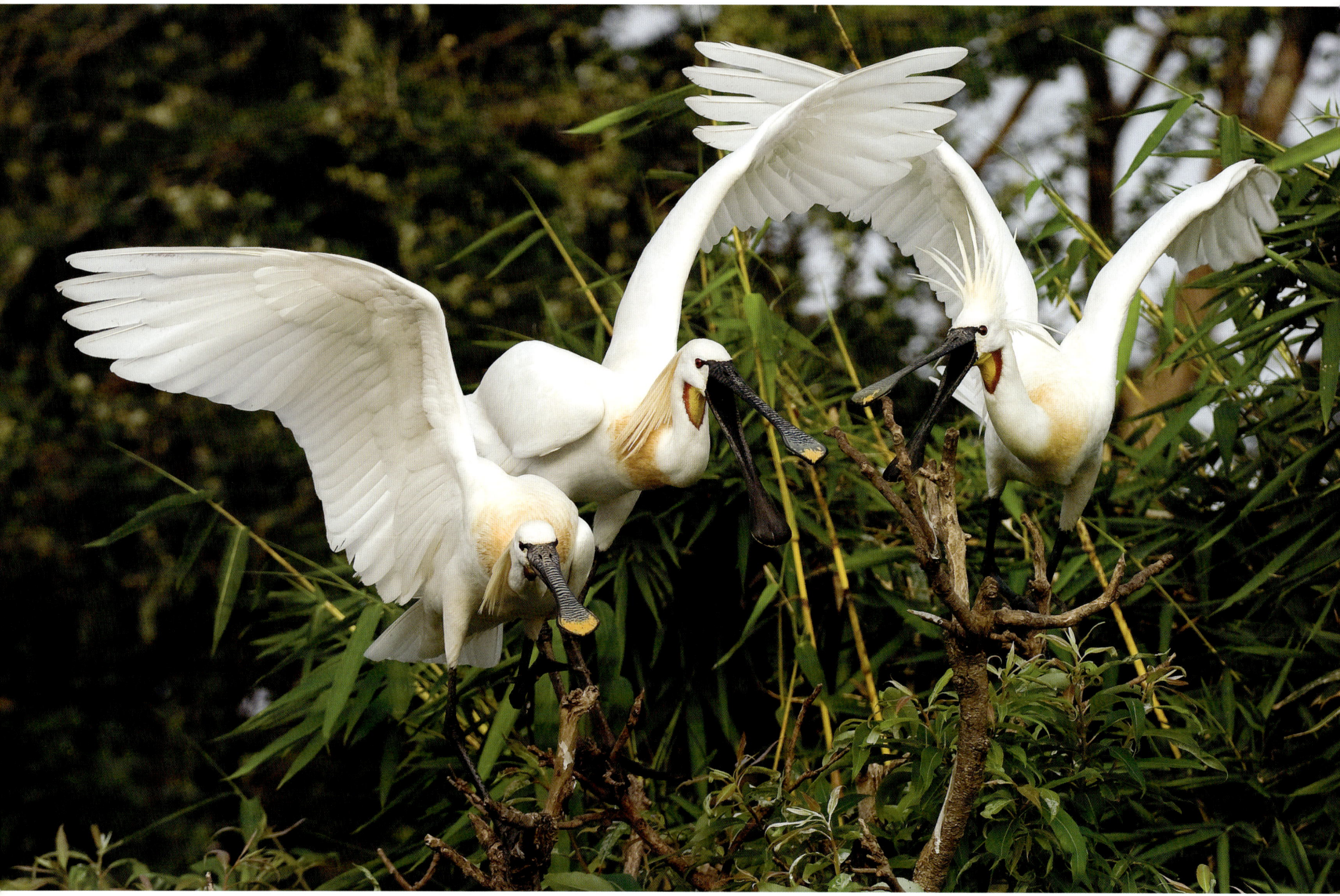

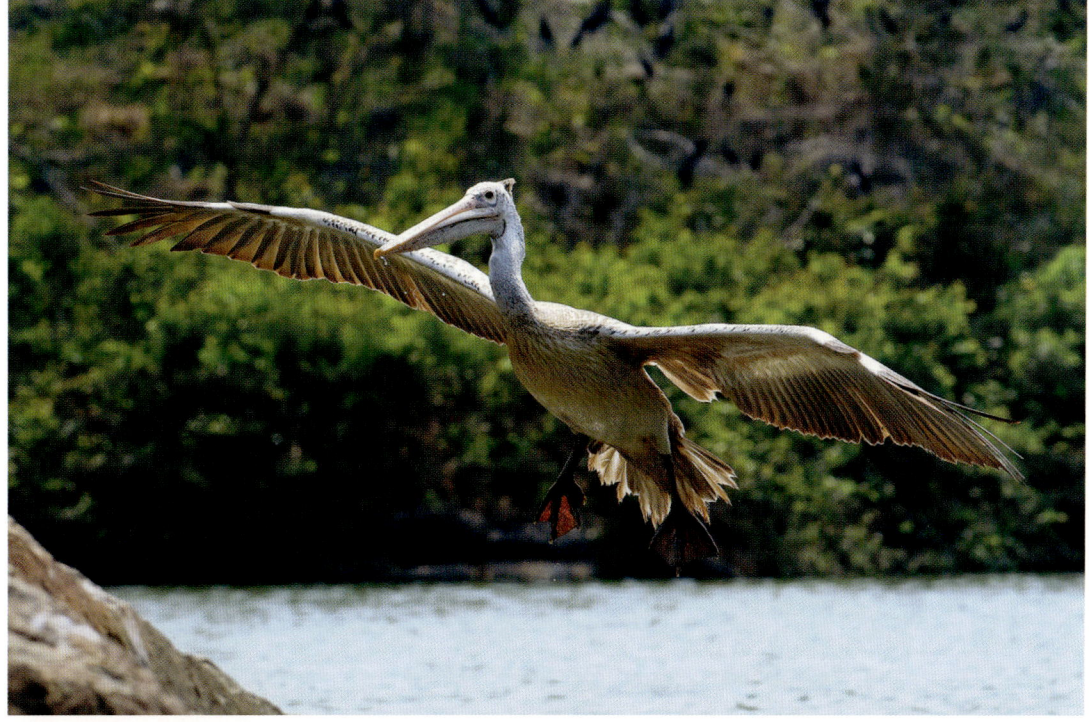

During the breeding season, the bird colonies on the Kaveri River are bustling with activity. While the Eurasian spoonbills are still squabbling over their choice of mate (left), spot-billed pelicans are already commuting to their nests (above right). The pied kingfisher (below right) builds a nesting cavity in a steep slope of the riverbank.

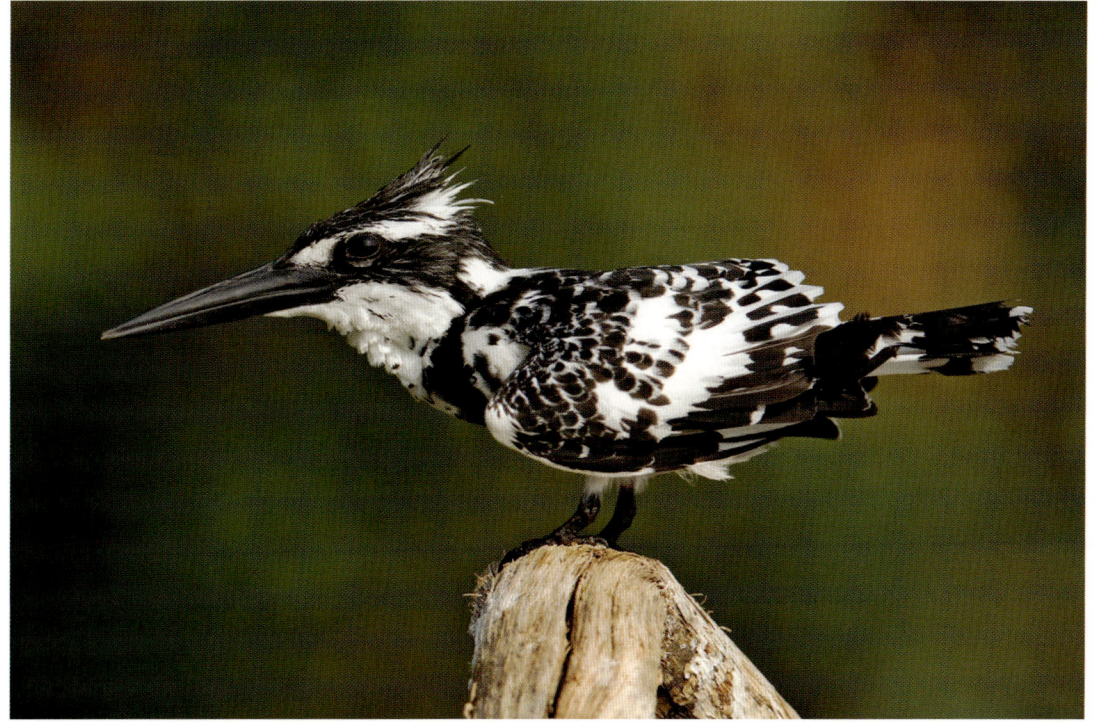

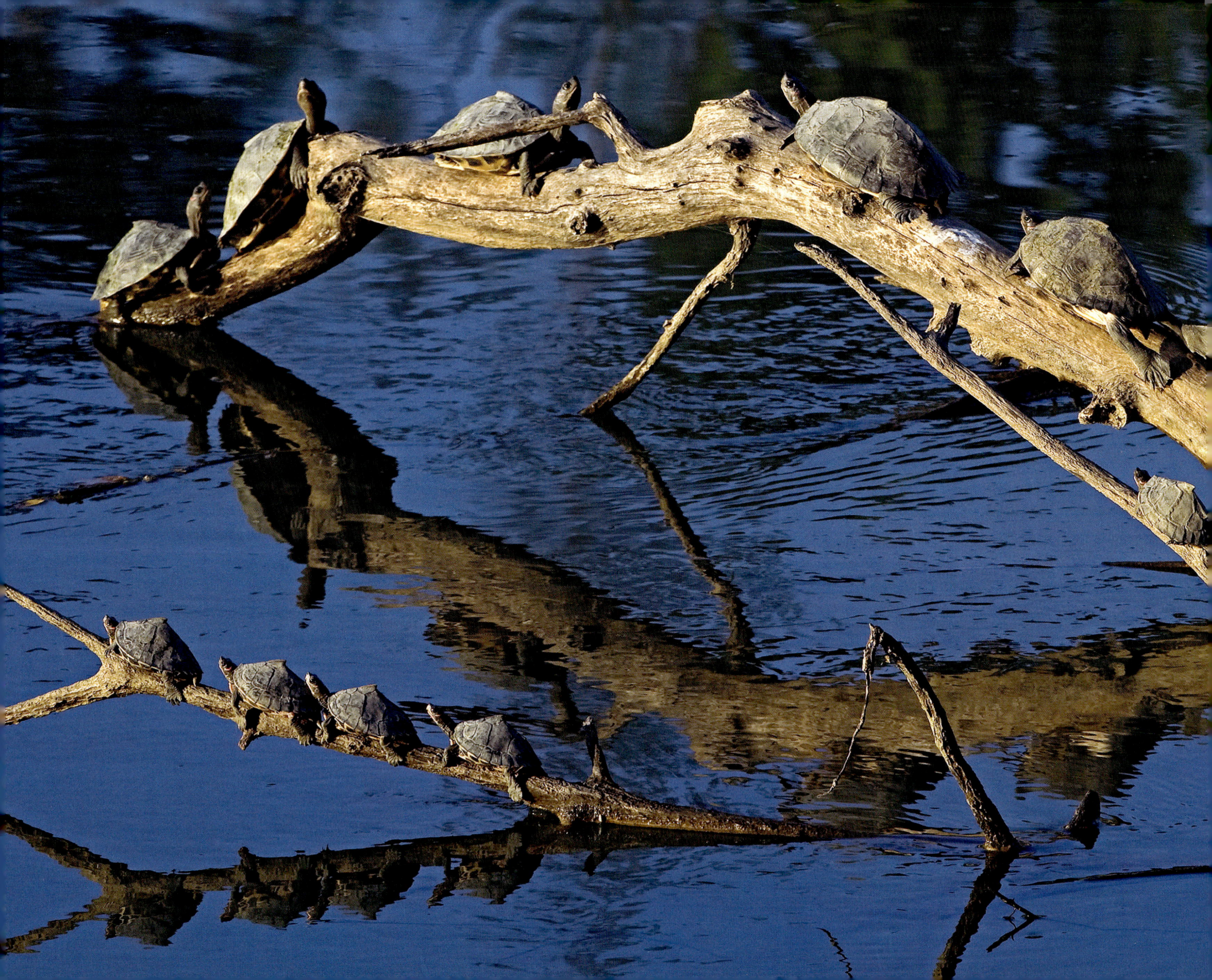

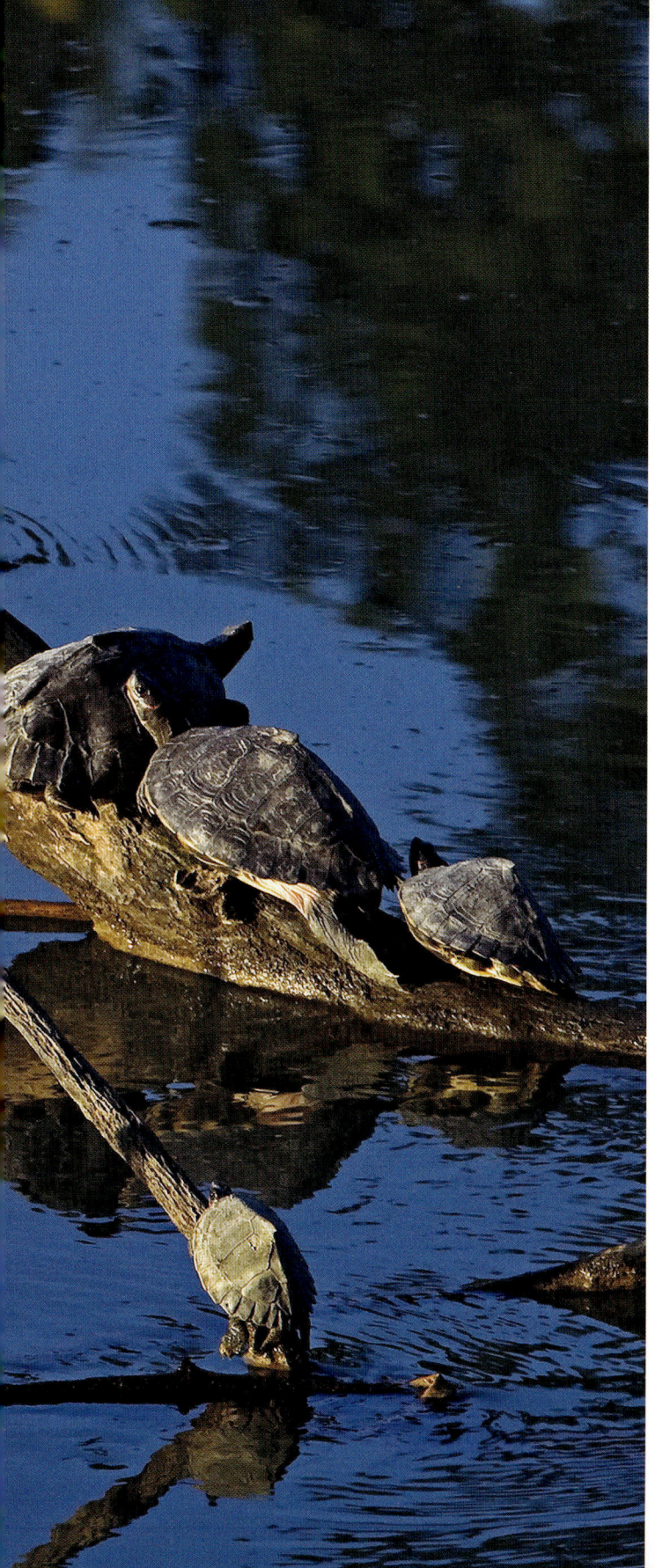

Assam roofed turtles have climbed onto a fallen log for their morning sunbathe. Like all reptiles, they are cold-blooded – their body temperature depends on the temperature of the environment. After a cool night they must therefore warm up, to reach their 'operating temperature'. Dragonflies are cold-blooded, too. Overnight they have assembled to sleep communally. As soon as the first rays of the sun have warmed these insects, they will be darting off in different directions.

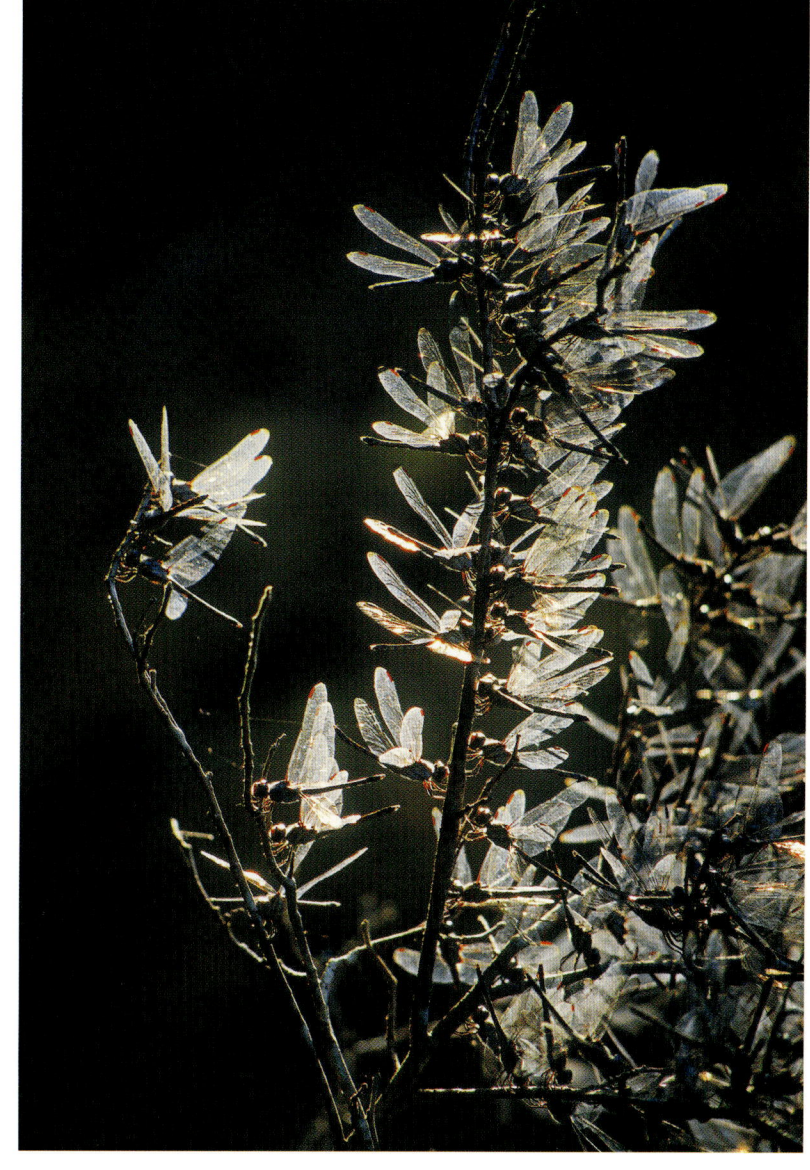

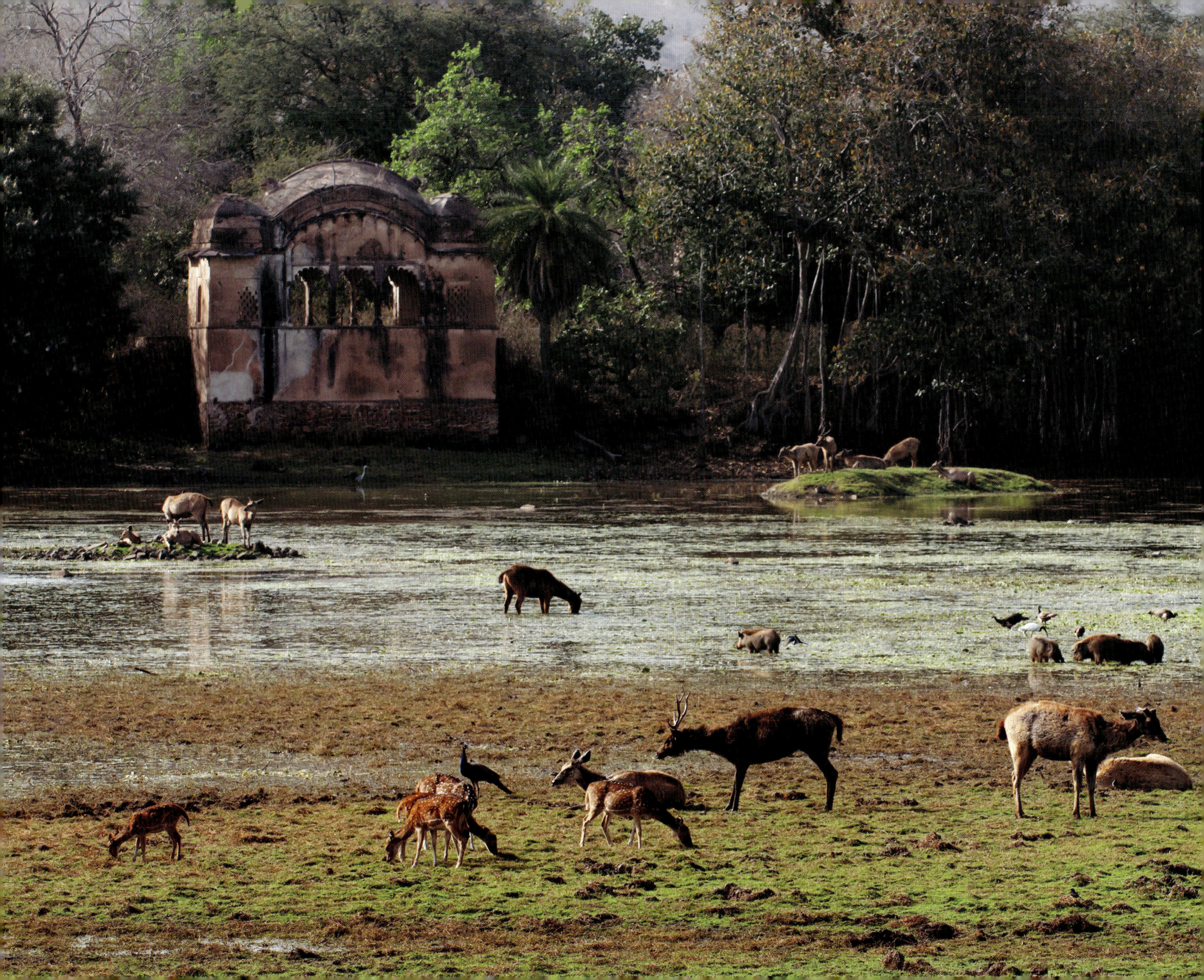

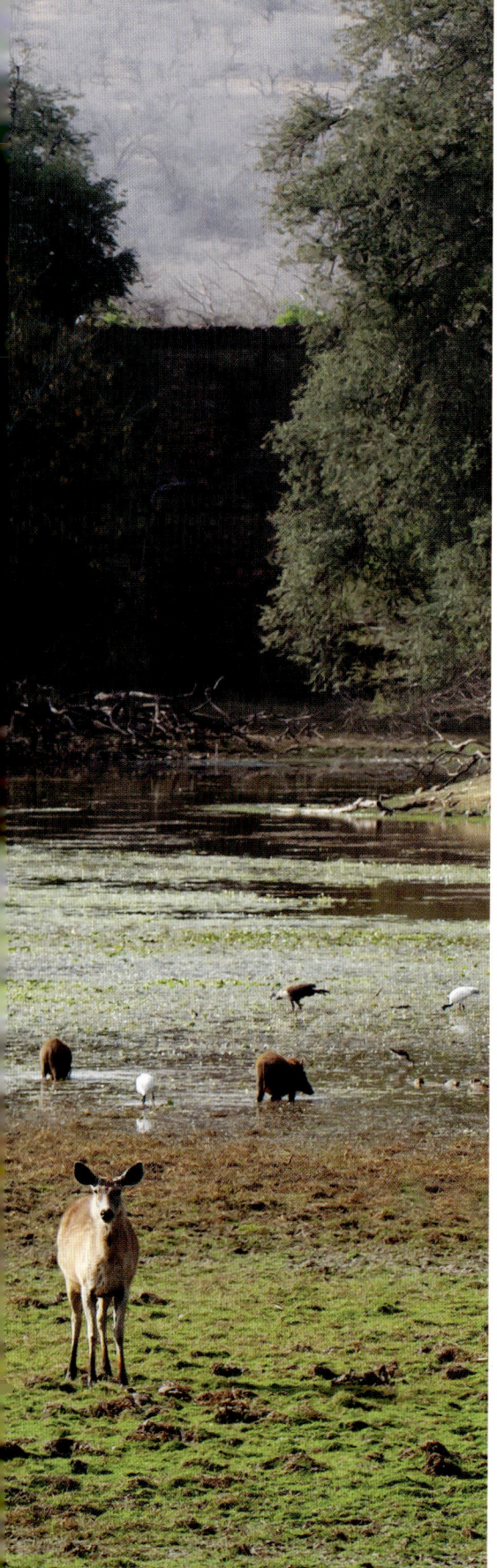

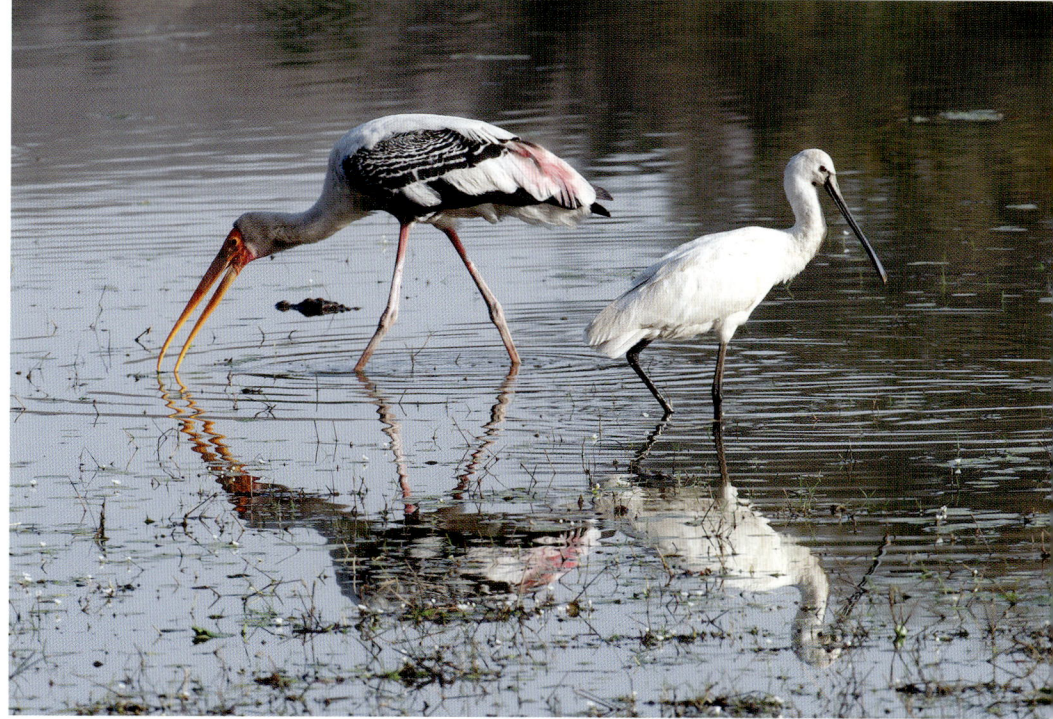

Ruins line the edge of Rajbagh Talao lake in Ranthambore National Park. Such water bodies attract lots of wild animals, as the surrounding areas are very dry for several months. Sambar deer, spotted deer, wild boar, peafowl and a great variety of other birds can often be seen foraging here. These waters are a nursery ground for marsh crocodiles as well, but these youngsters (right) do not yet pose a threat to the water birds.

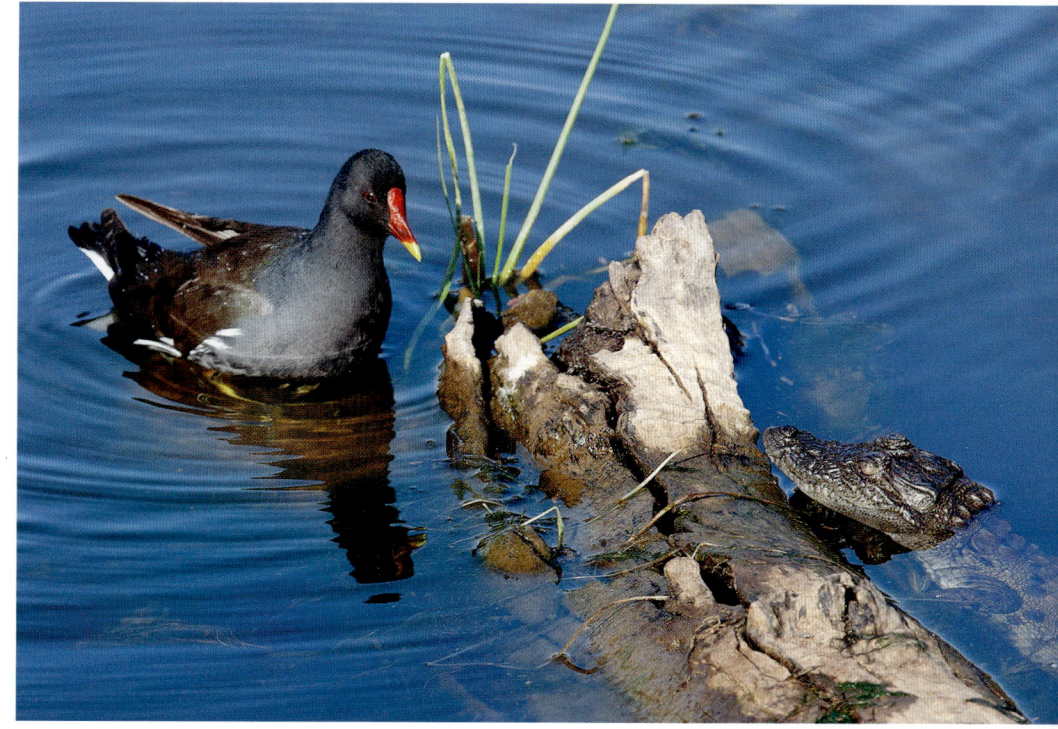

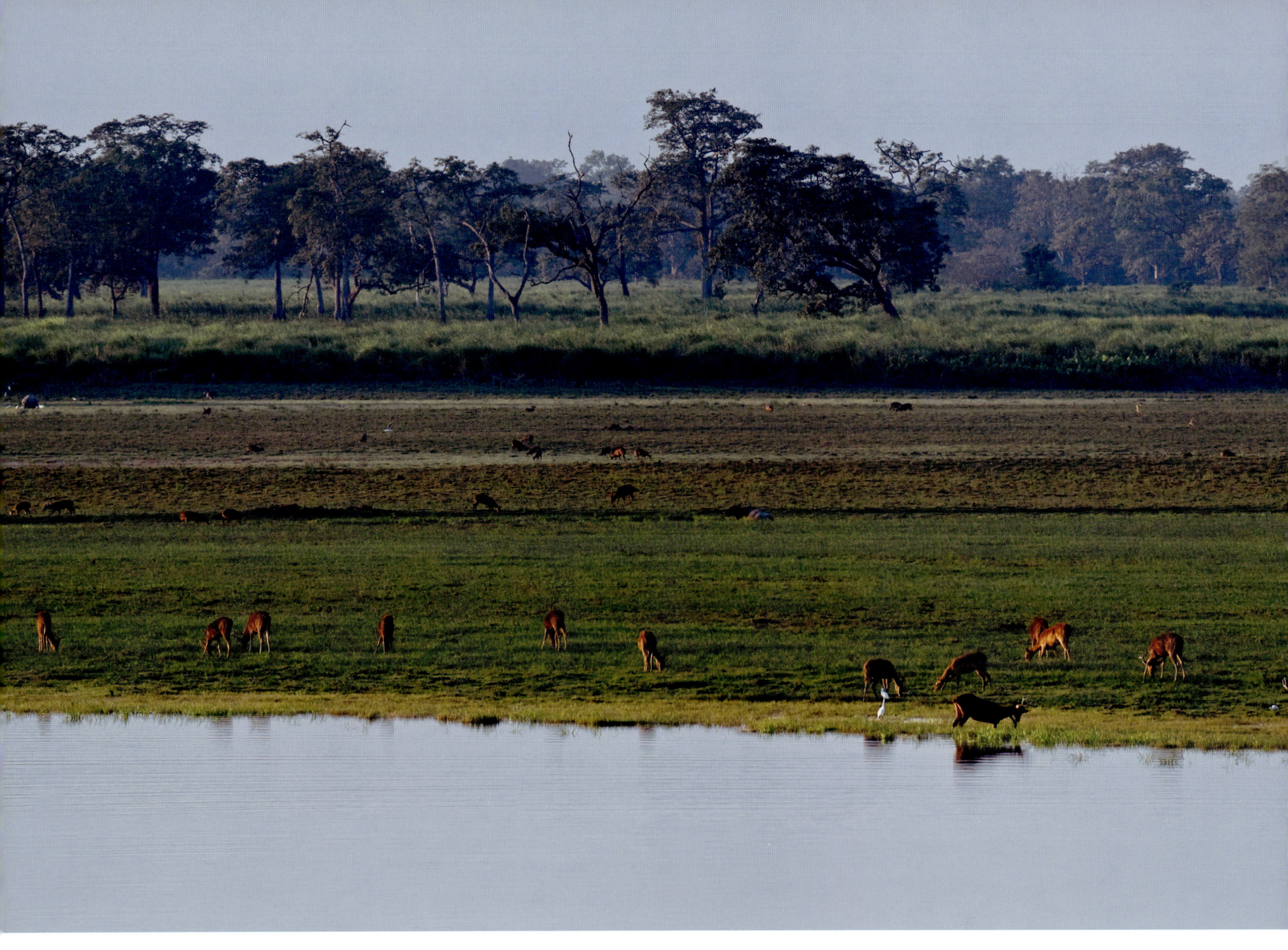

An Indian rhinoceros, Asian wild buffalos, swamp deer and hog deer graze on a floodplain of the Brahmaputra River in Kaziranga National Park in the Indian state of Assam.

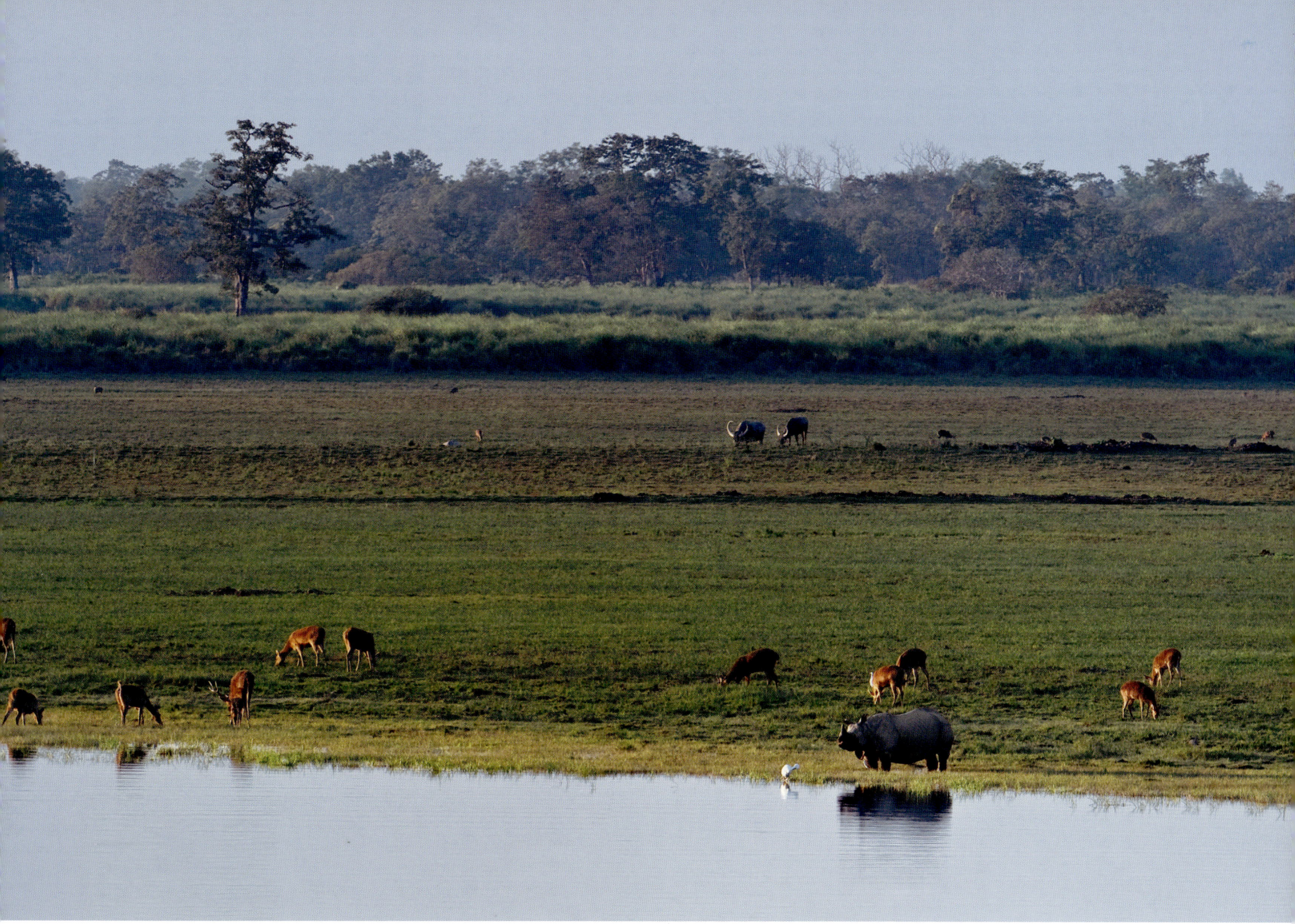

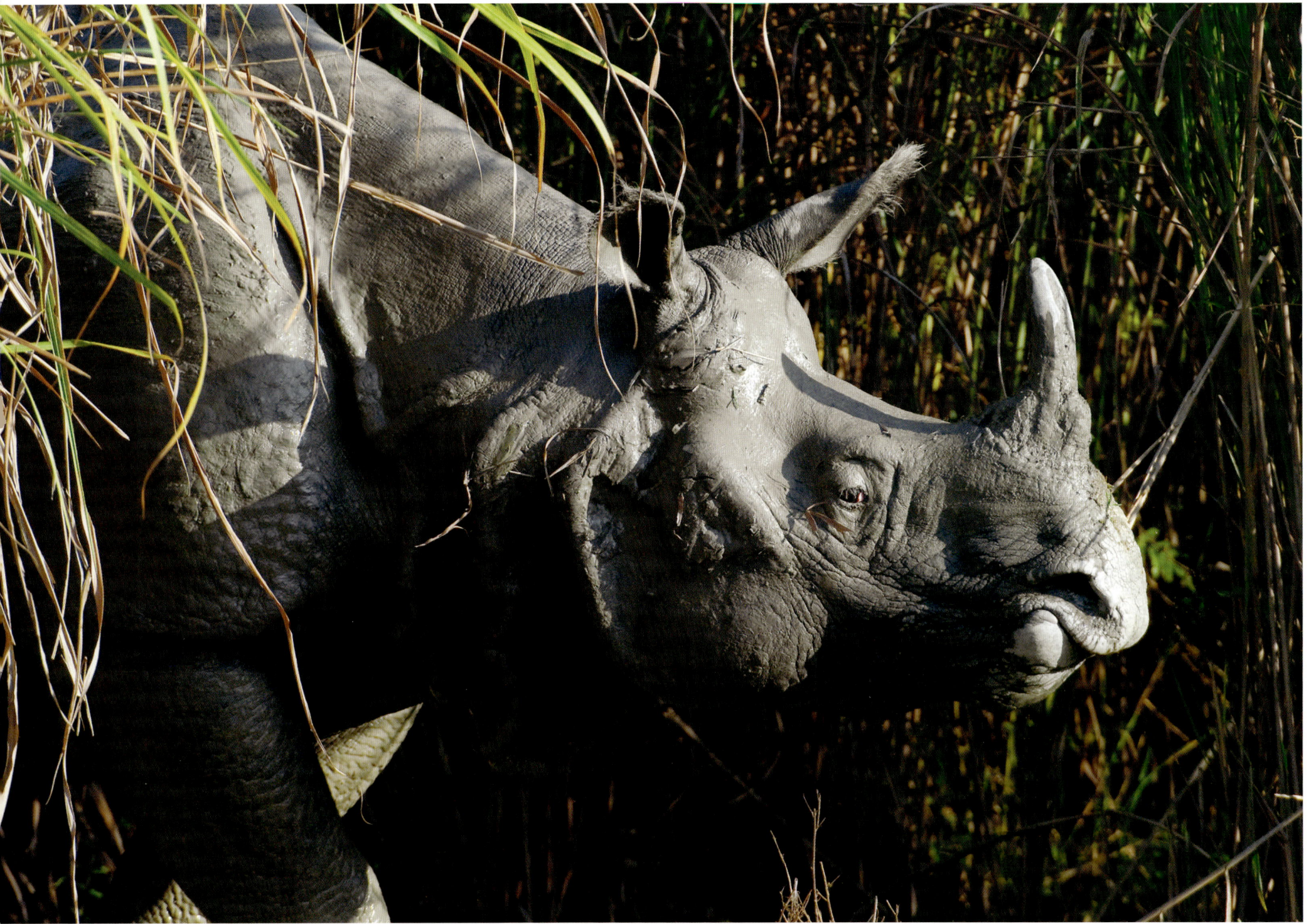

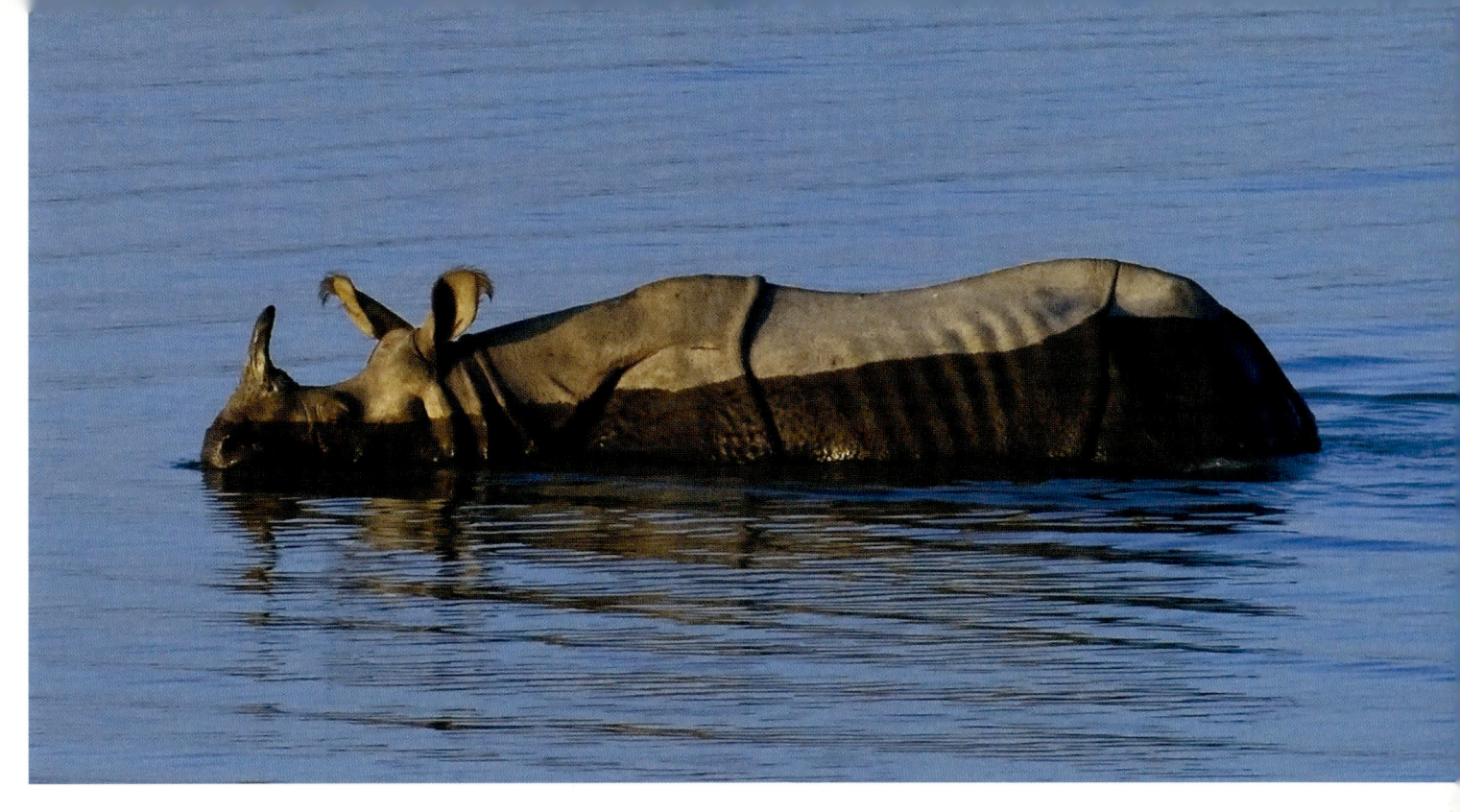

With its prehistoric appearance, the Indian or greater one-horned rhinoceros looks like a relic from a bygone era. The species is adapted to living in marshland, where it moves easily between water and land. About a hundred years ago these rhinos were on the verge of extinction, and their survival is one of conservation's great success stories. Kaziranga National Park played a key role in protecting these animals. Today, the total wild population of Indian rhinoceroses stands at approximately 3,300 individuals – about 70 per cent of them live in Kaziranga.

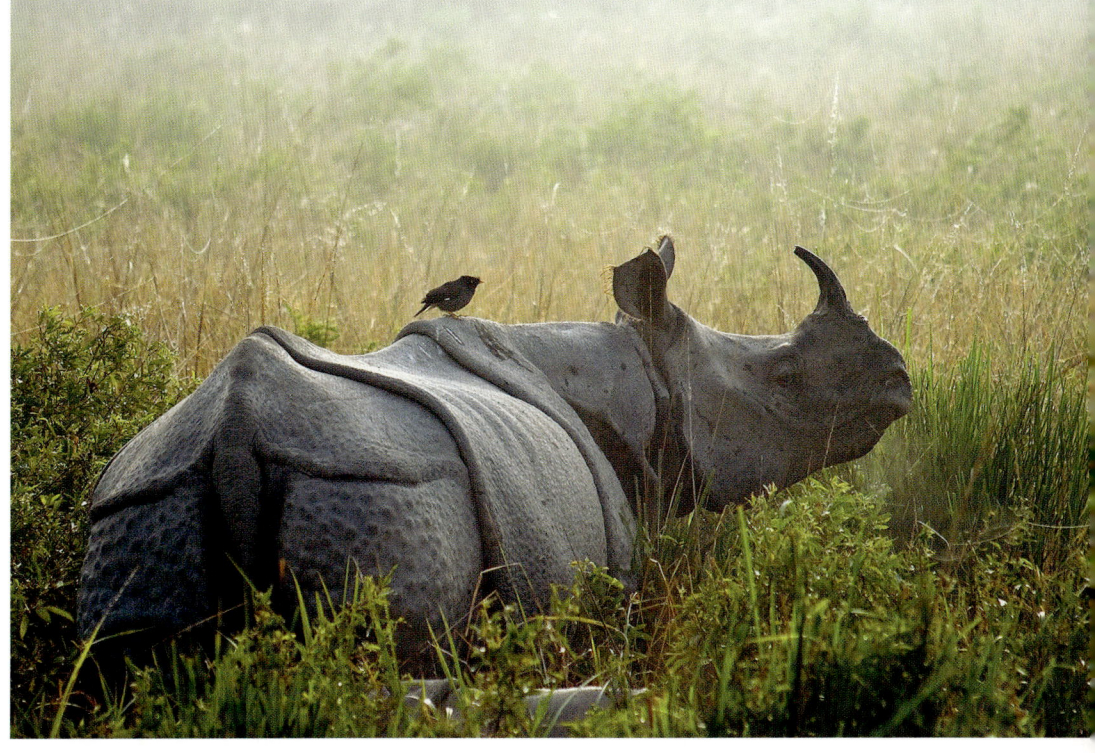

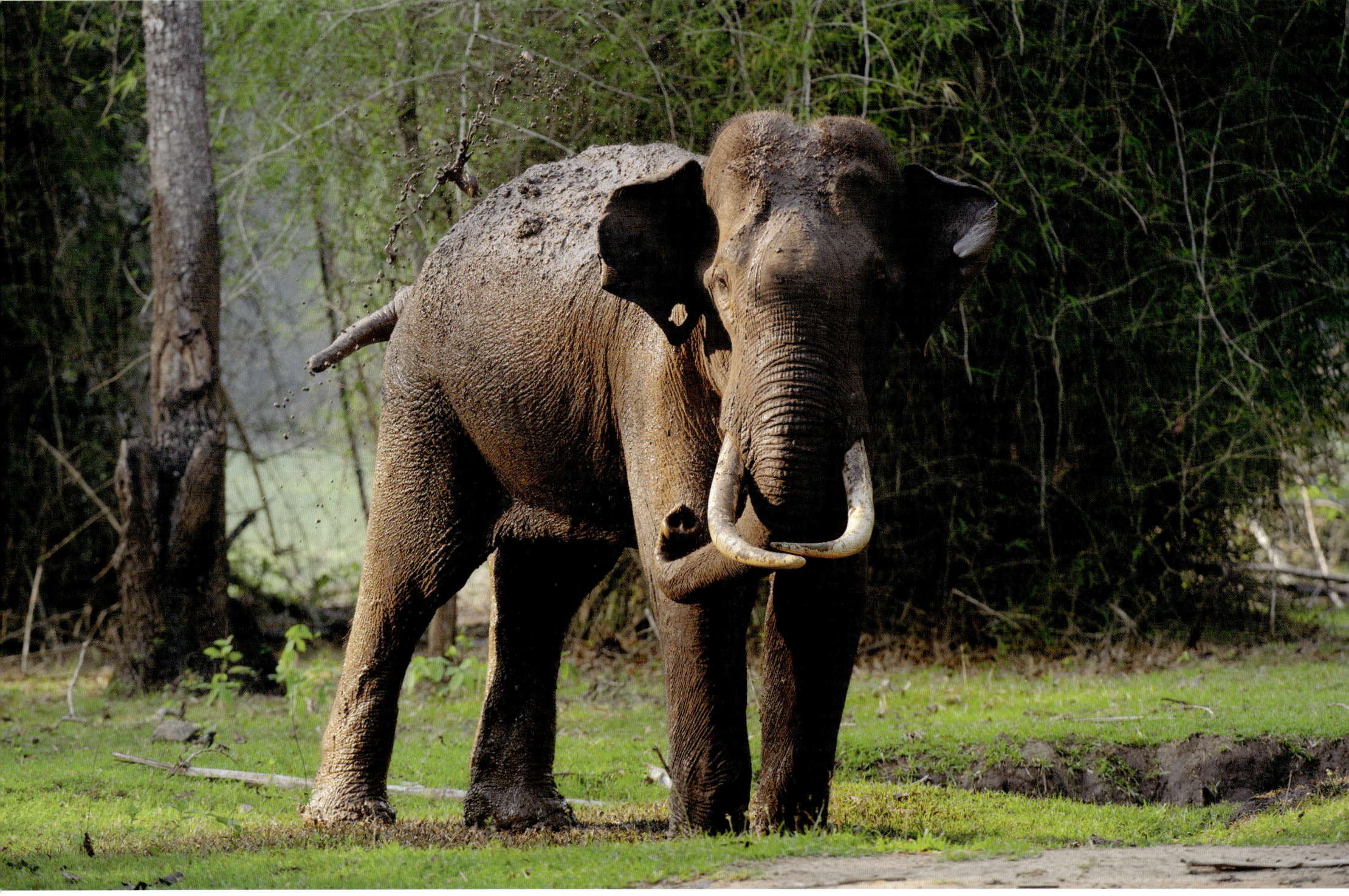

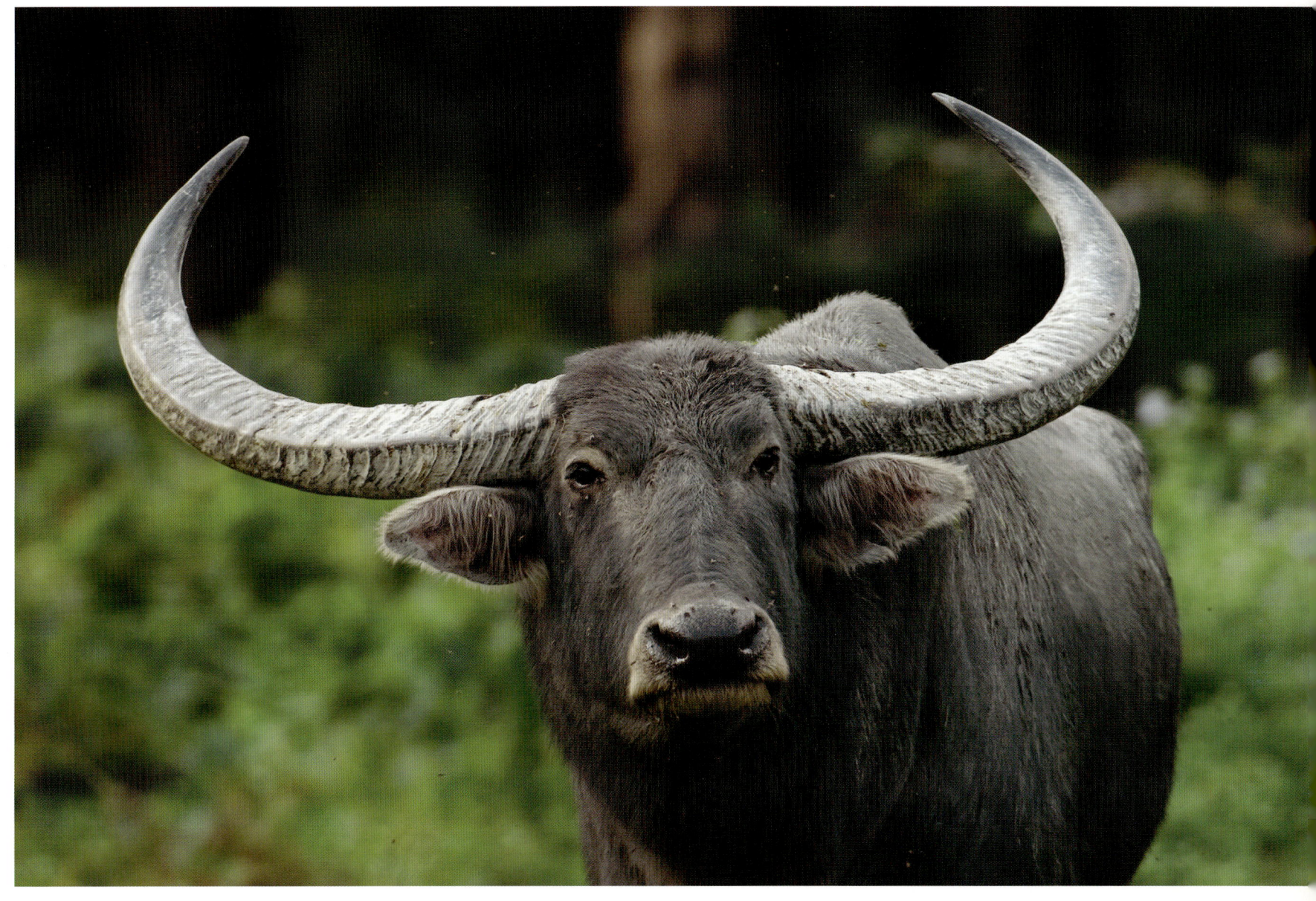

Kaziranga National Park is home to the Asian wild buffalo, the ancestor of the domestic water buffalo. The buffalo's characteristic massive horns are an impressive sight. Asian elephants live in the same area, and enjoy taking a mud bath.

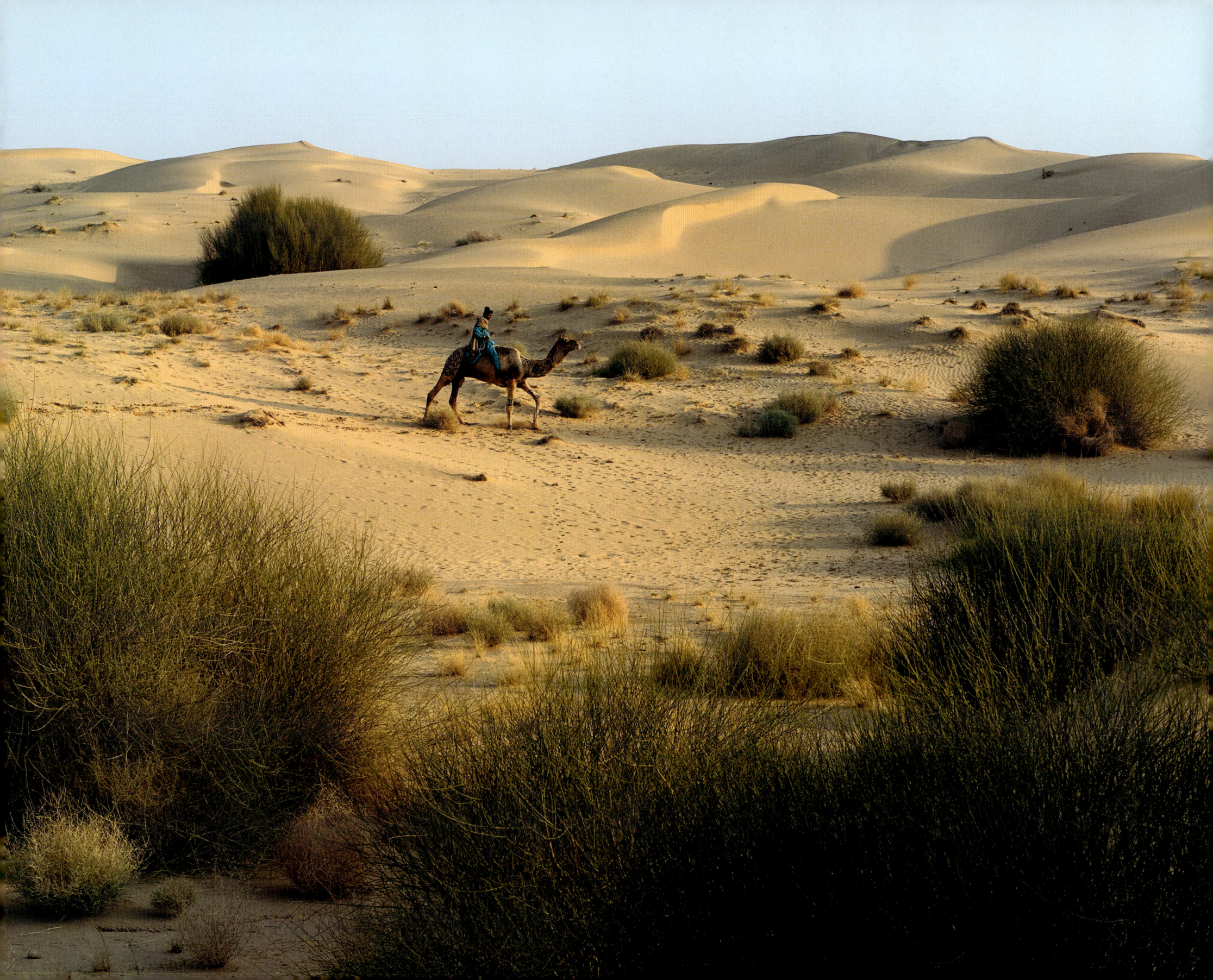

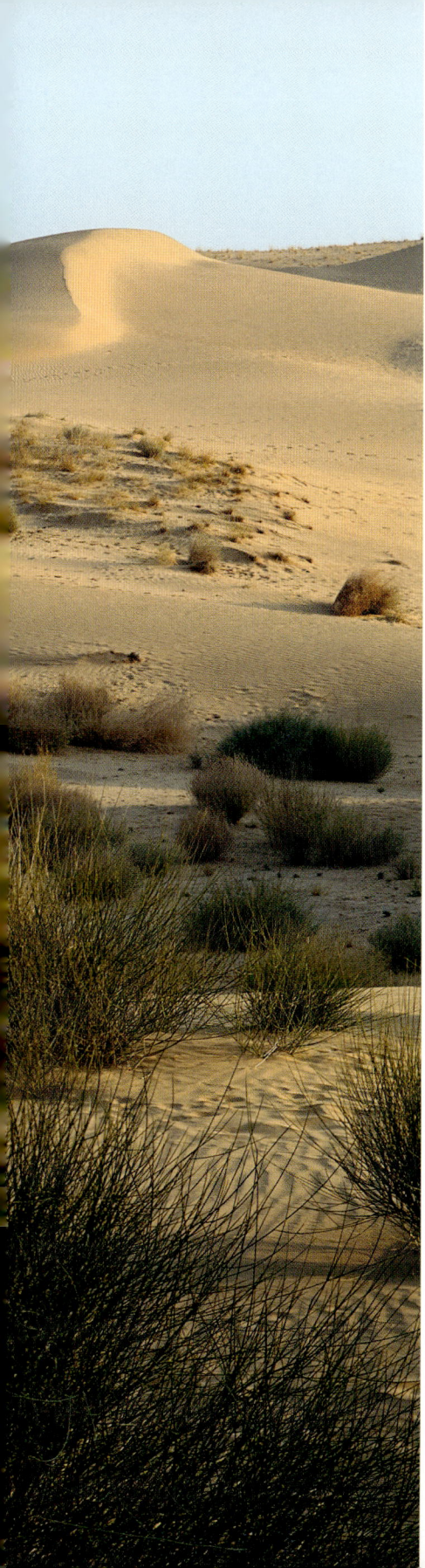

The Magic of the Desert

India

Northwestern India is especially hot and dry. The Thar Desert spans this part of the country – a region of extraordinary beauty with a rich cultural heritage.

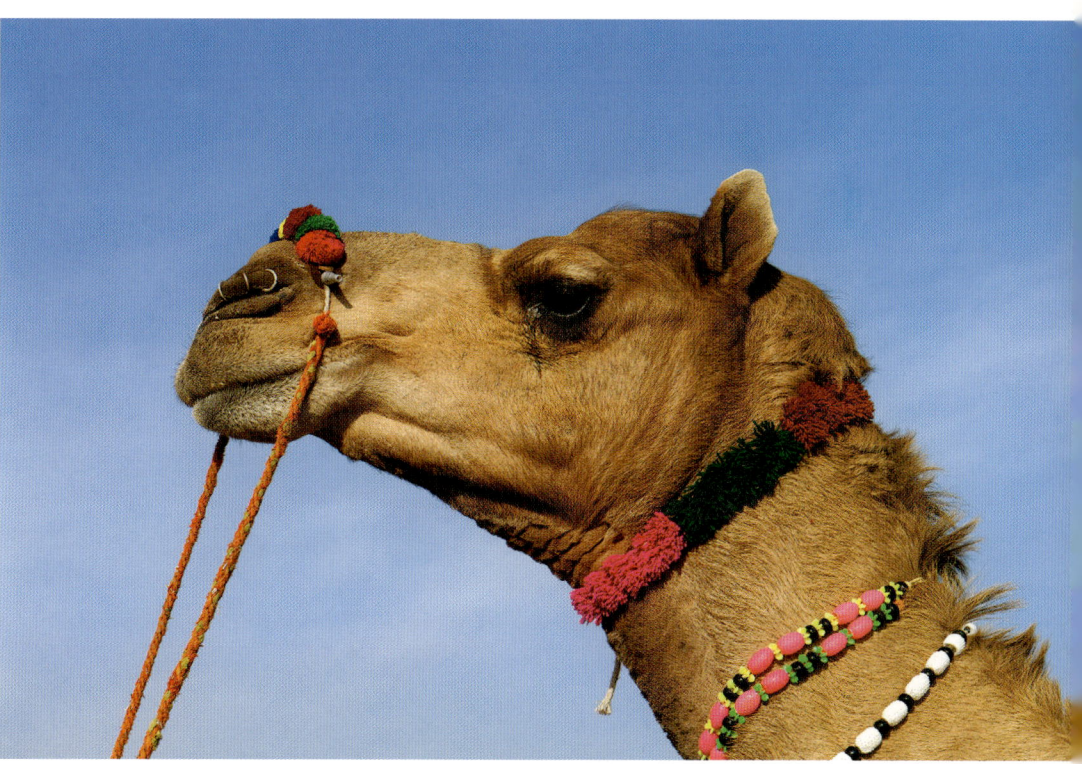

A Journey into the Past

The constant, monotonous bleating of the camels cuts through the early-morning haze. They're hungry – just like their owners. The camel fair at Nagaur in Rajasthan slowly comes to life. Bleary eyes squint in the morning sun. Everywhere figures emerge from their bedrolls and shake out dusty blankets. Here and there men rekindle their campfires to dispel the lingering night chill. The smell of burning wood spreads, accompanied by the scent of fresh tea. I hear tin cups striking against kettles. Simple flatbread is being prepared over the fire. At last the camels get their breakfast too. An old man spreads out a sack of grain in front of them. Several animals converge on it at once, bending their long necks down and consuming the fodder together.

Virtually nothing in this scene recalls the present day. No cars, no power lines – modern technology is nowhere to be seen. The hustle and bustle of the fair is probably just as it was several centuries ago. It is as if I have travelled back in time to the Middle Ages – simply by taking a long flight from Europe.

In this world of sand and dust the humble dromedary is an important beast of burden. It is cheaper than a car and can be taken cross-country. It traverses sand dunes with ease, where automobiles would quickly get stuck. It can tolerate large temperature fluctuations and can survive for extended periods without water. No wonder trade in the animals is flourishing. Tens of thousands of dromedaries change hands at the fair, along with smaller numbers of cows and horses.

To attract customers, the animals are artfully decorated with various devices. Colourful braided cords adorn the heads and necks of the camels. Ornamental painting usually begins on the animal's face, but it can extend over the flanks as far as the legs. There are even camel hairdressers, who laboriously trim the coat into decorative patterns. The owners present their animals with pride. One of them offered me his ship-of-the-desert for the equivalent of about 500 euros.

That's a sum that few farmers can afford to pay. As the cost of diesel and other fuels rises on the world market, the demand for dromedaries increases as well, because they are cheaper to maintain than vehicles. Yet traditional animal husbandry is a difficult undertaking in the Thar Desert. Even today, many people live simple lives in mud huts. The younger generations, especially, aspire to prosperity and happiness in the city – and go to Jaisalmer.

This desert citadel has long been famous for its wealth. It lies in the western part of the Thar Desert, close to the Pakistani border on an ancient trade route. Its strategic position between India and Central Asia was the key to its importance. Camel caravans brought a flourishing trade to the 'Golden City', as Jaisalmer is also known. Ornate dwellings arose alongside splendid temples and an impressive fortress. Numerous historical sites in India have been destroyed over the course of history and survive only as ruins – but ancient Jaisalmer is inhabited to this day, and looks for all the world like a movie set. Elaborately adorned women in multi-coloured saris stroll through the winding lanes in a scene reminiscent of a tale from *One Thousand and One Nights*. Here in Jaisalmer it can still be found – the magic of the desert.

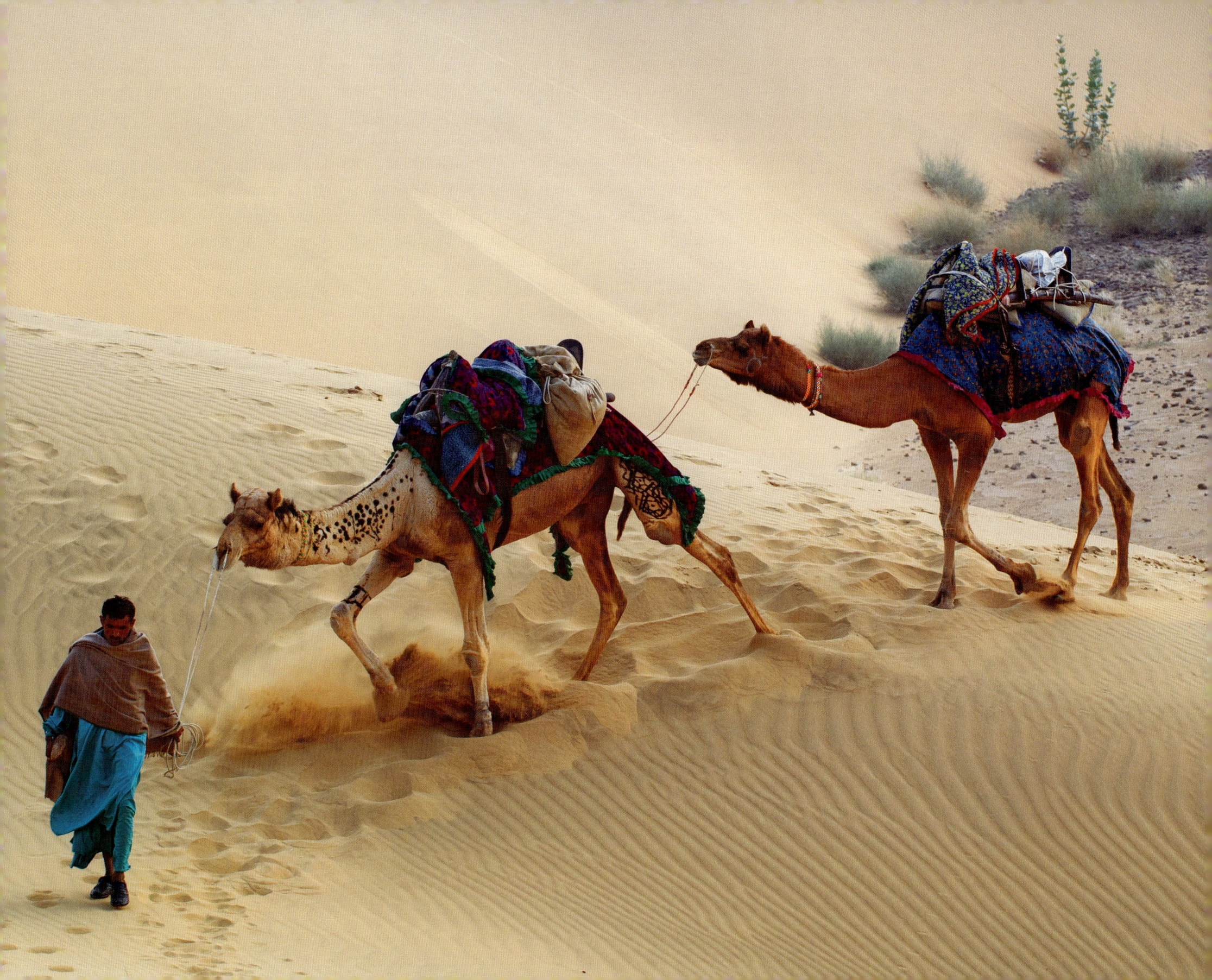

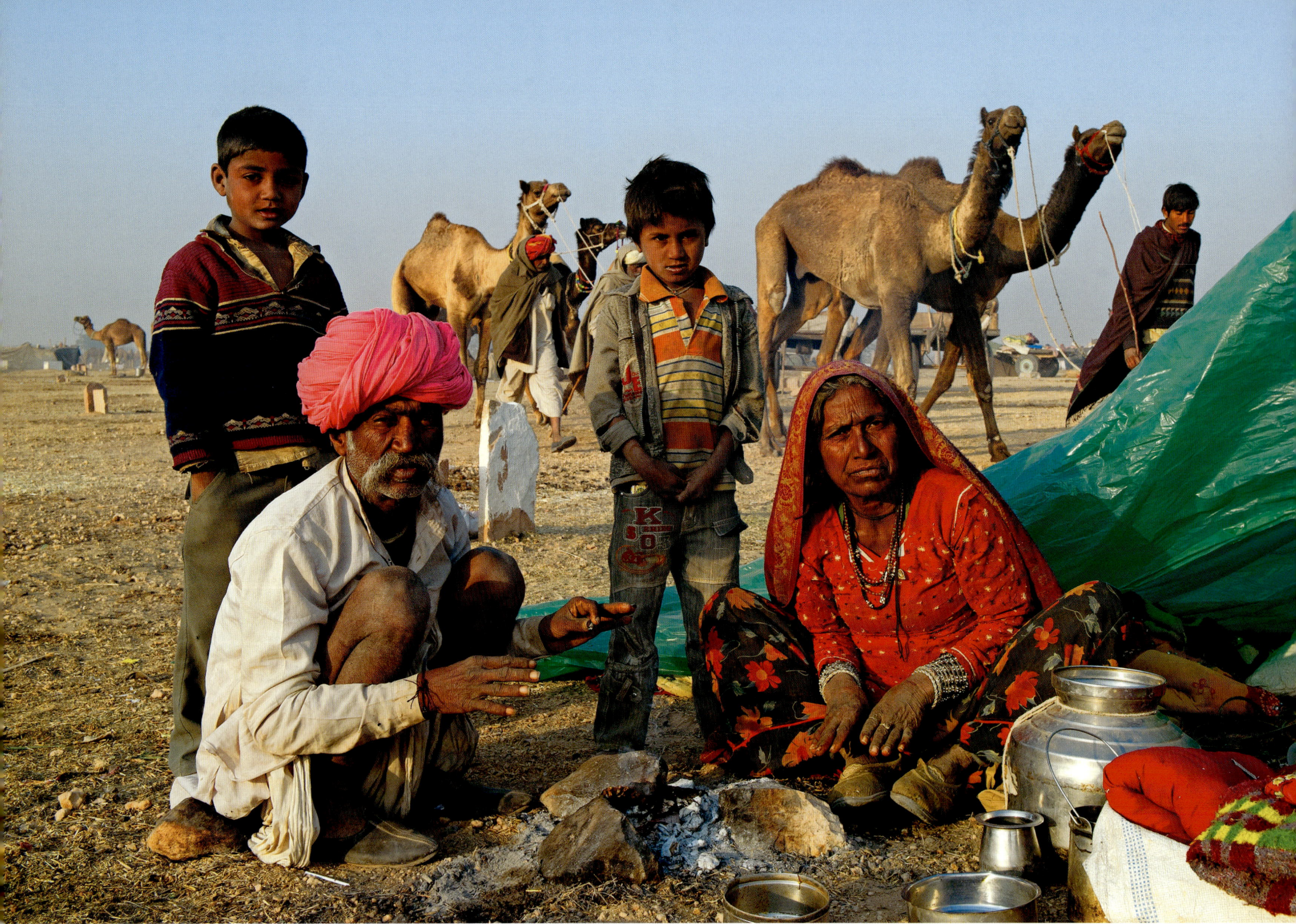

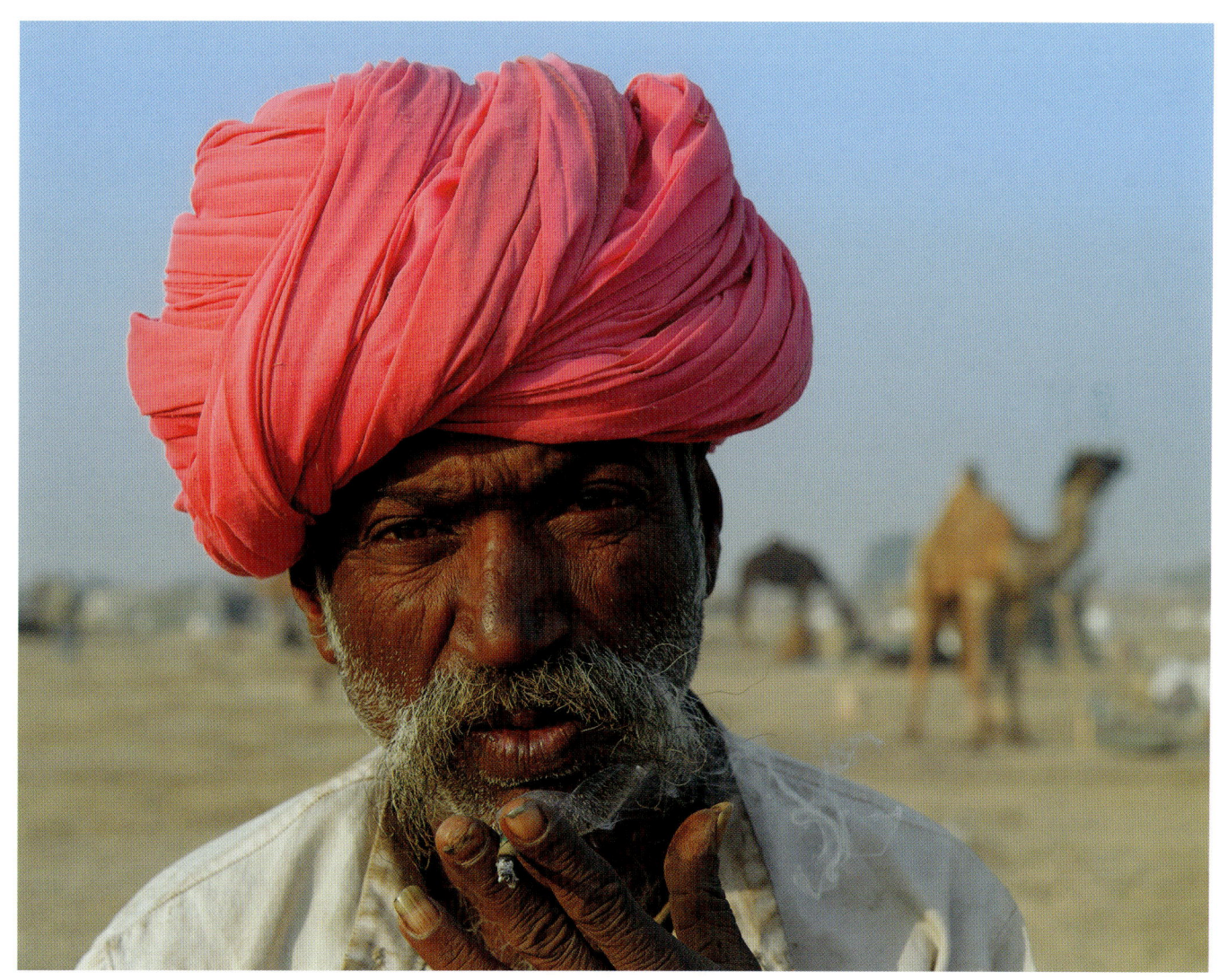

The camel fair in the city of Nagaur in Rajasthan attracts traders and buyers from all over the Thar Desert. The event lasts for several days, and the whole family takes part. The wise old men of these clans enjoy a great deal of respect.

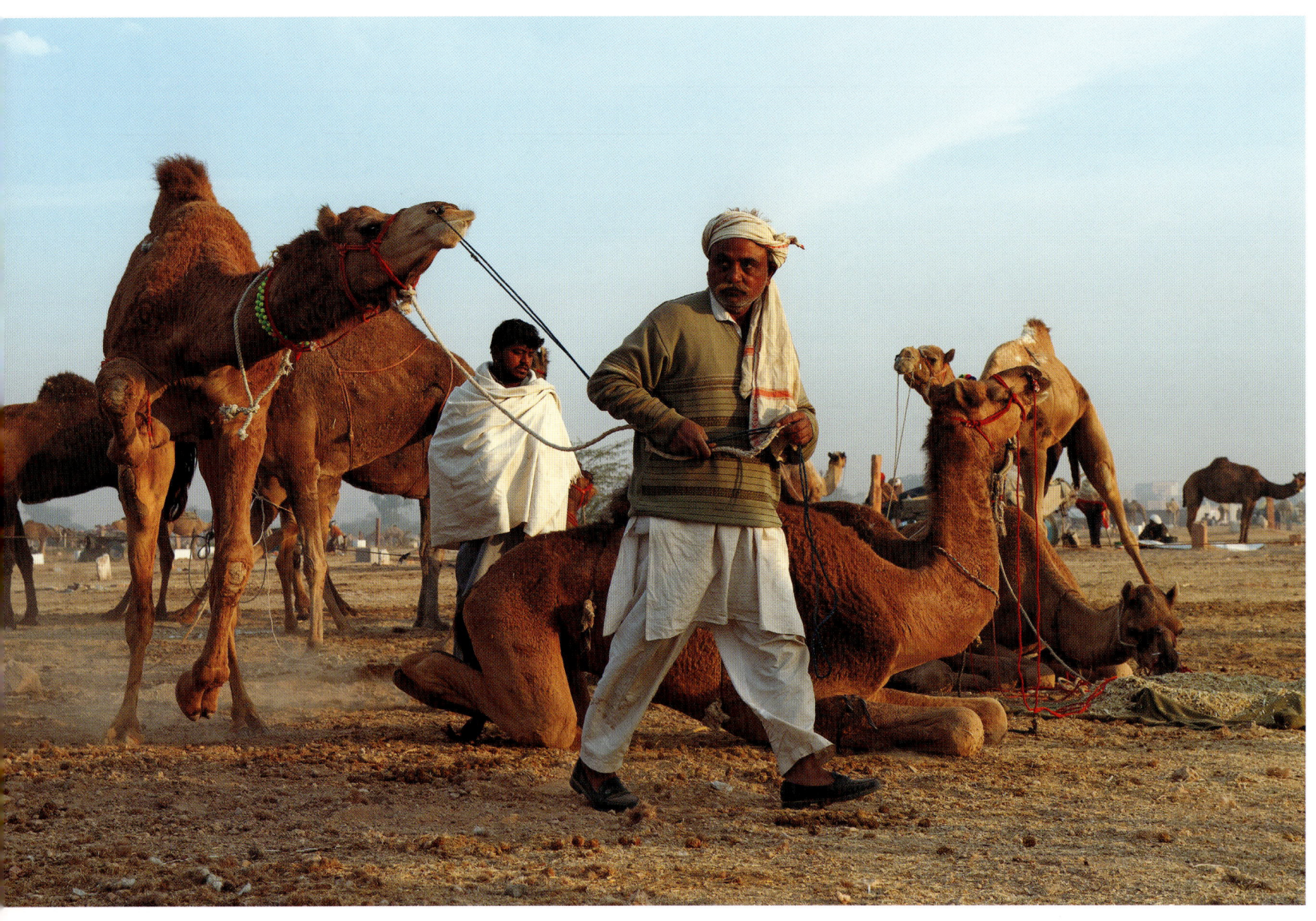

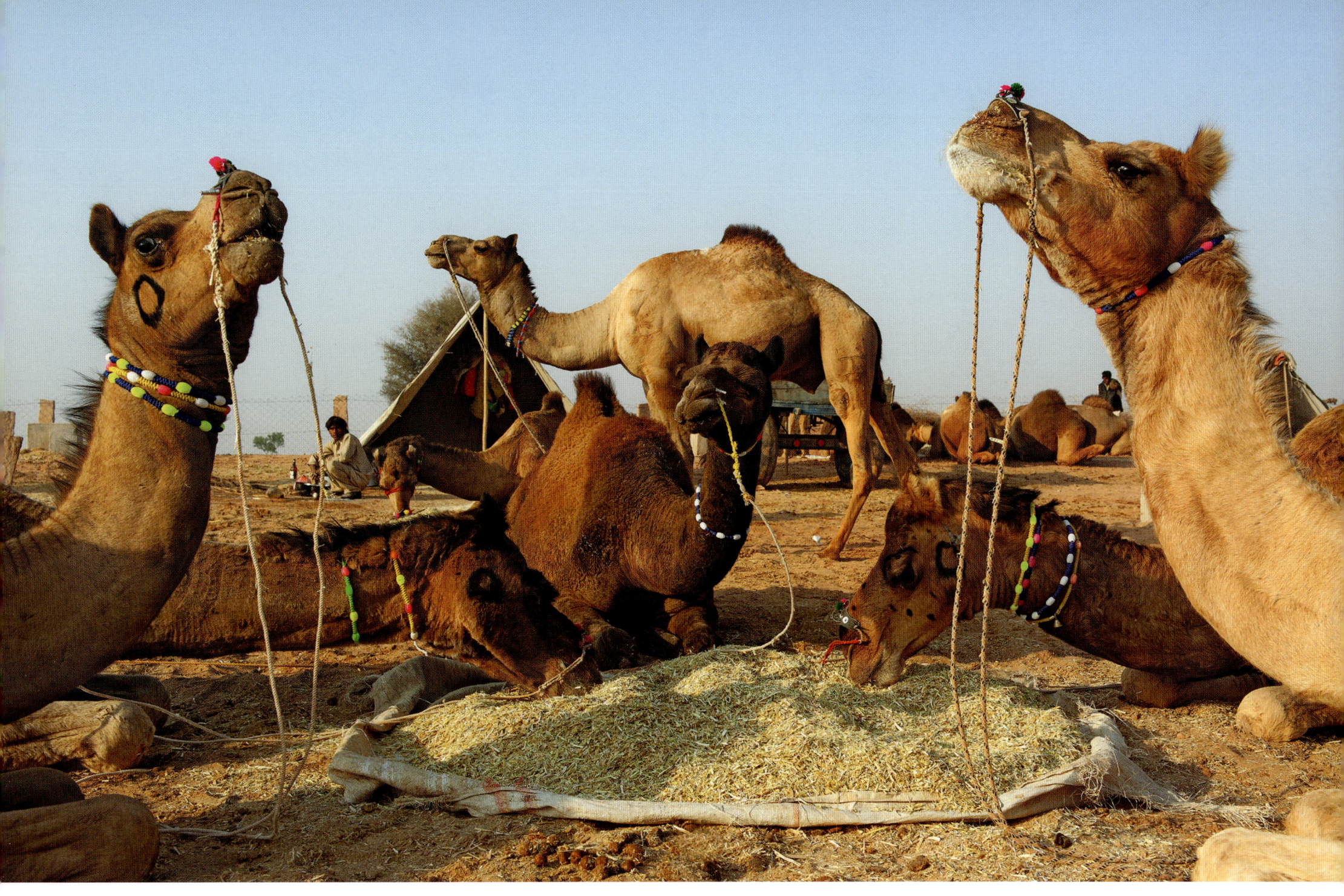

In the morning, the camels are herded together in small groups
for their breakfast, which is a sack of grain.

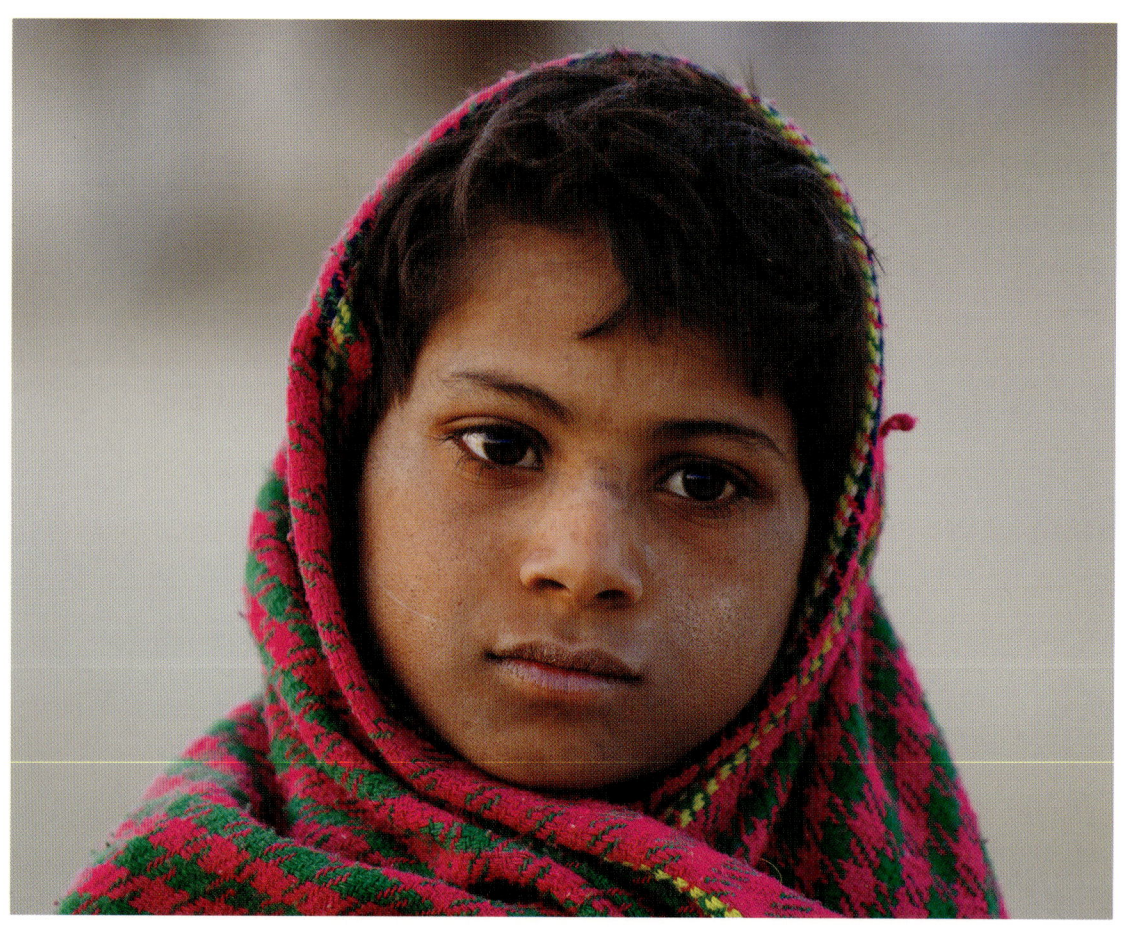

The people of the Thar Desert are widely scattered. When they gather in one location, on the occasion of the camel fair, it is a social event as well – and for the young women the desert festival functions as a marriage market. The boys support the family by guarding the livestock.

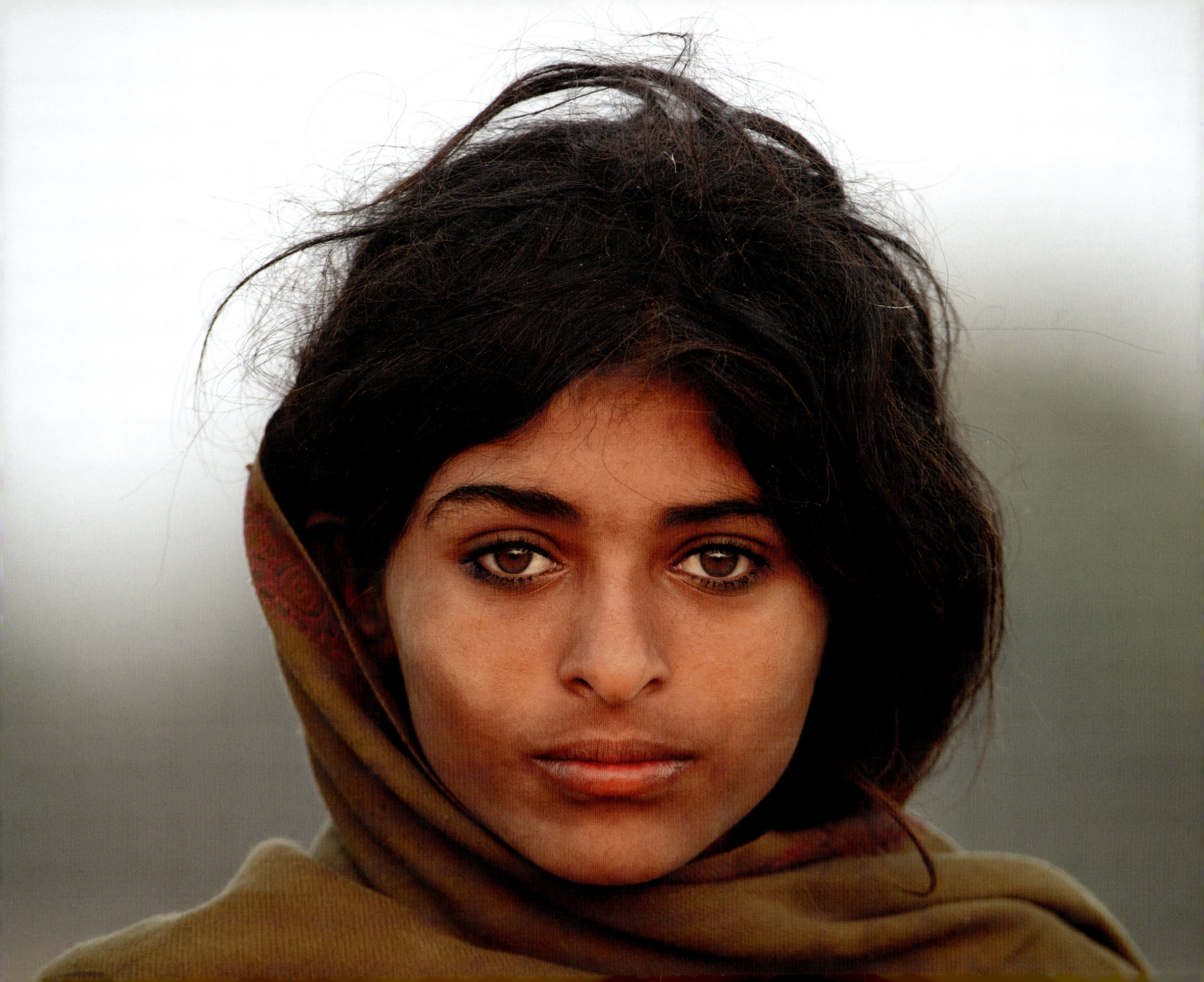

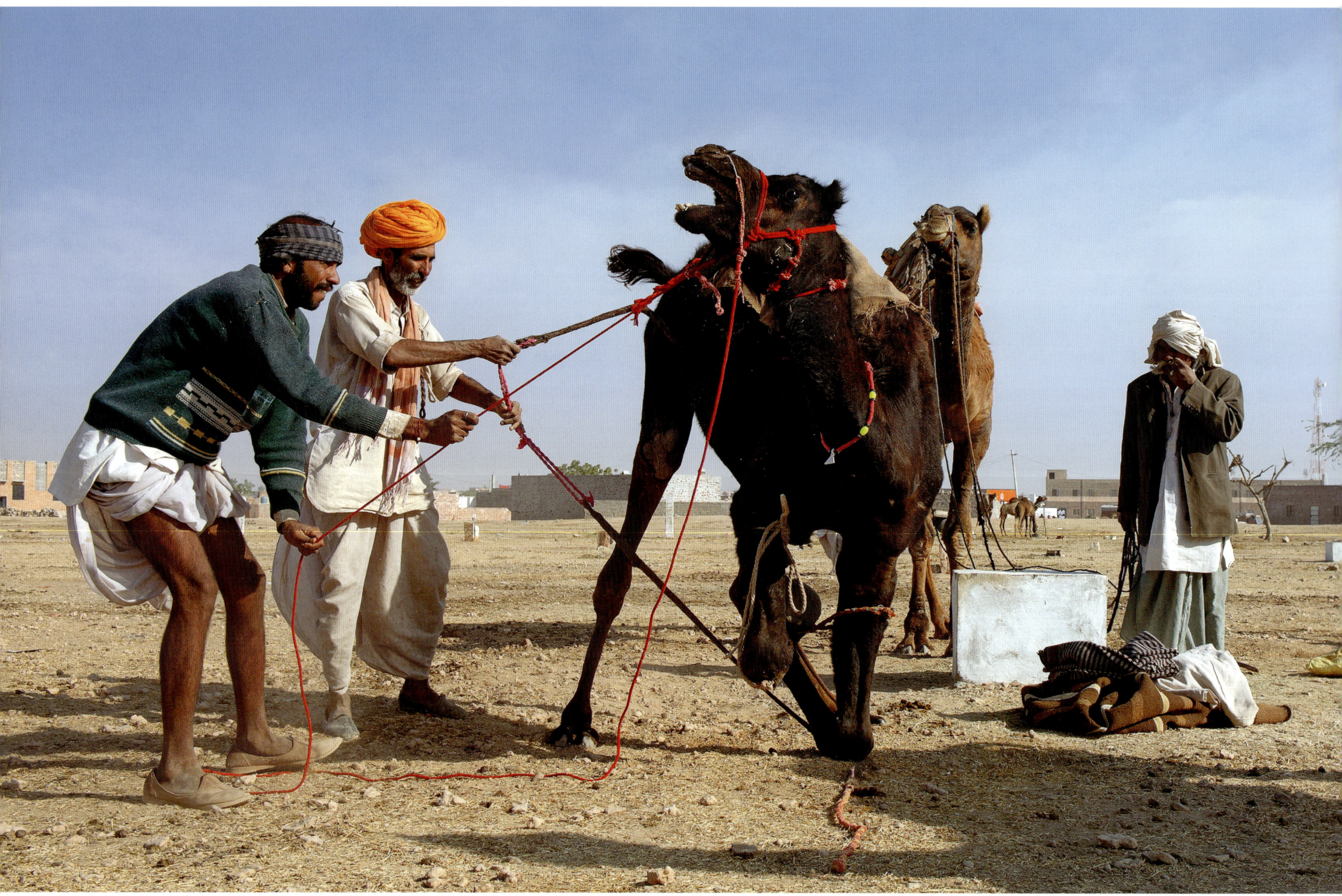

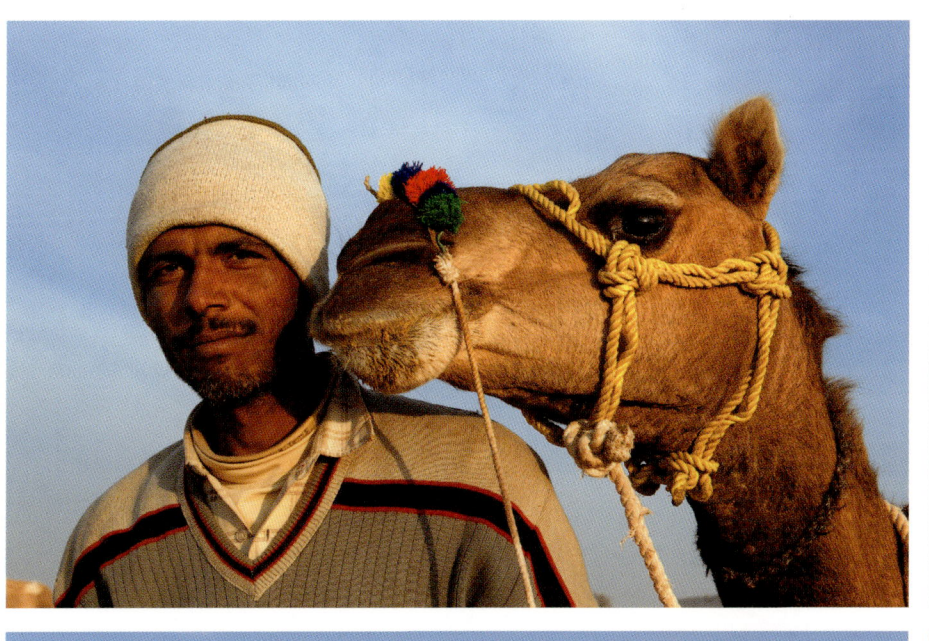
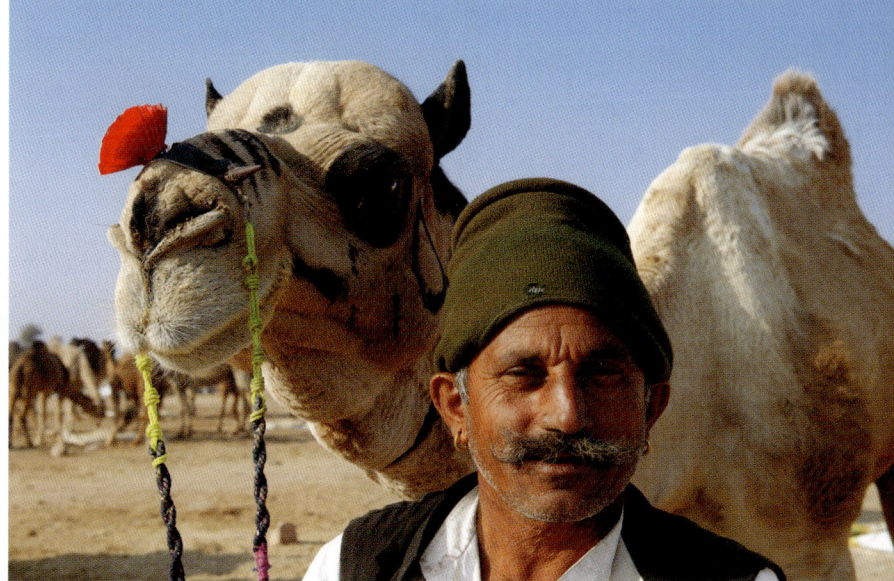
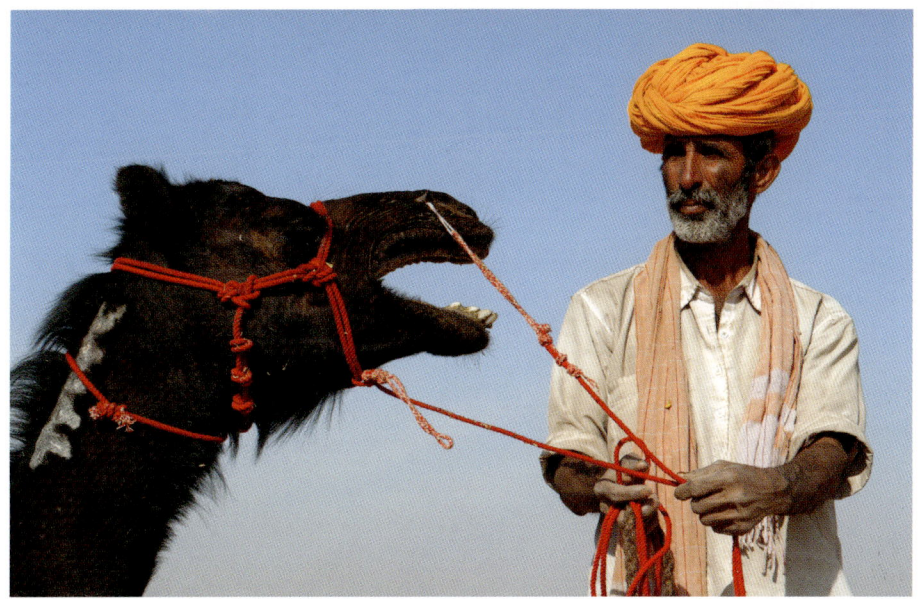
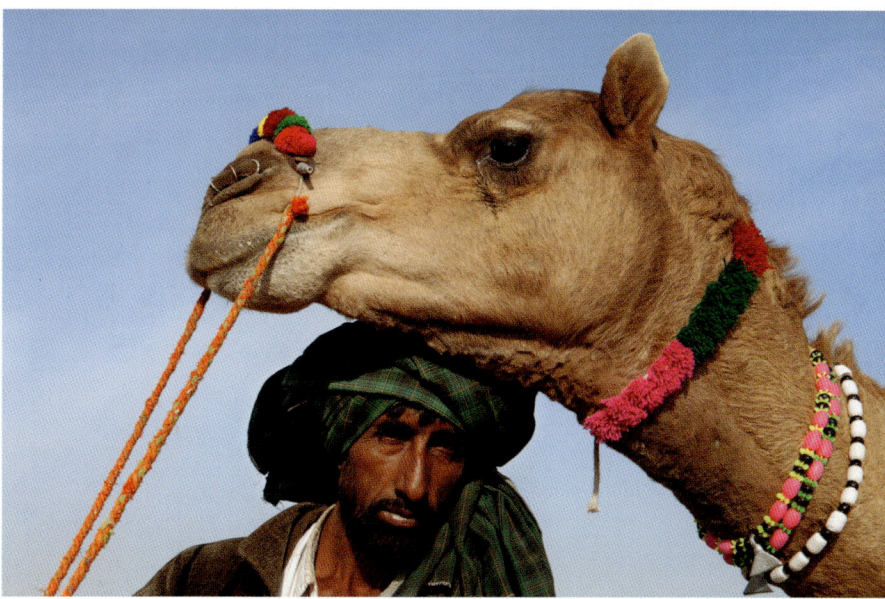

In the desert, dromedaries are indispensable as working animals, and their owners show them off proudly. But when the animals are sold and handed over to unfamiliar people, they sometimes show the stubborn side of their character (left).

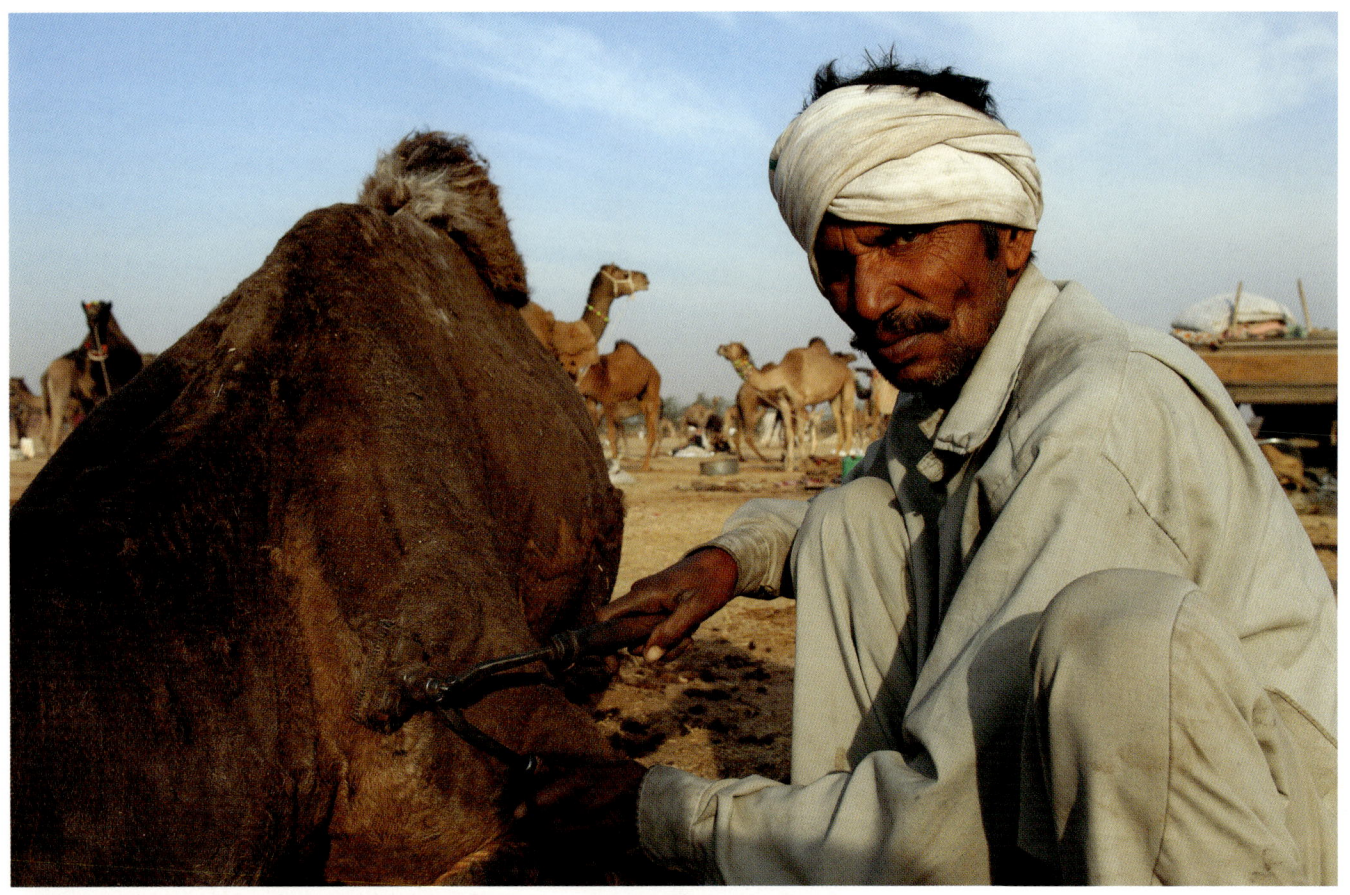

For the camel fair the animals are often elaborately painted. The finishing touch is then provided by a camel hairdresser, who cuts decorative patterns in the animal's fur.

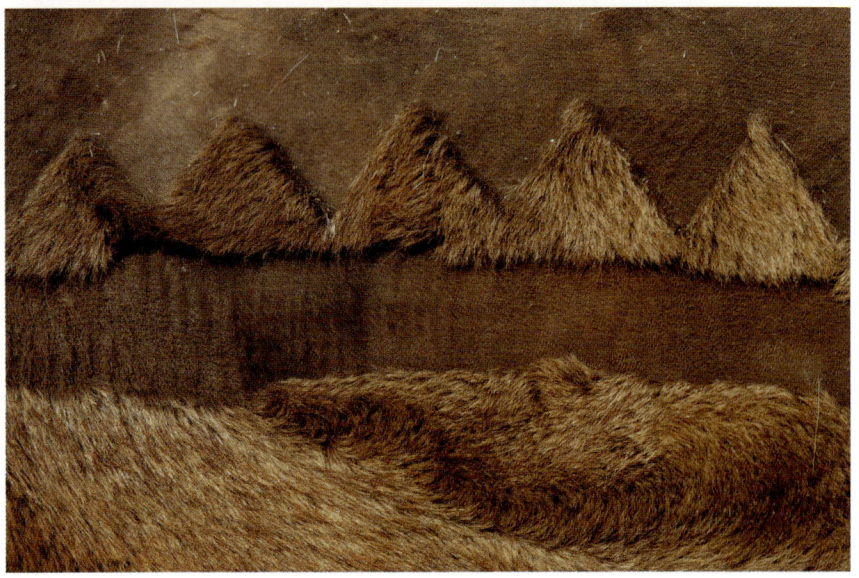

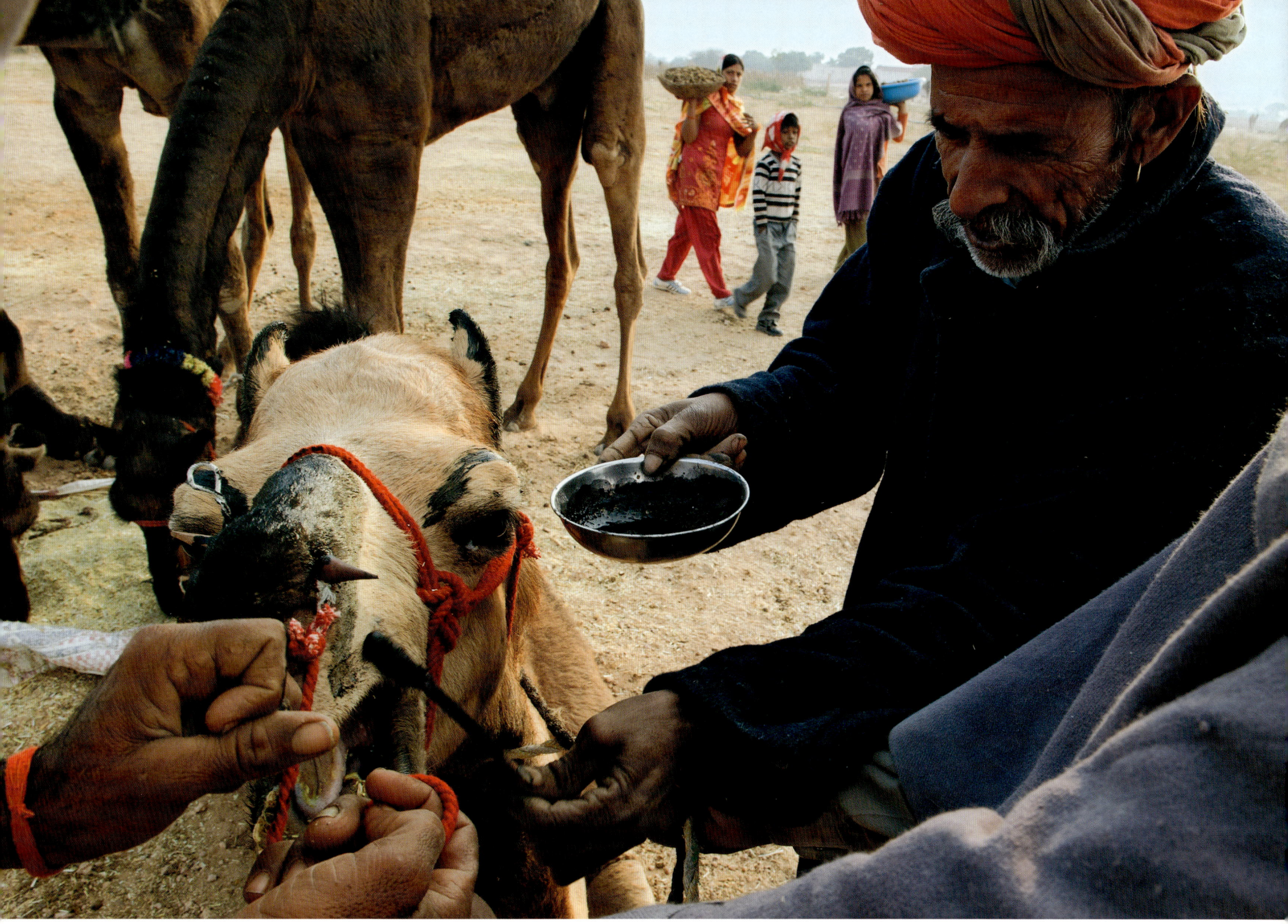

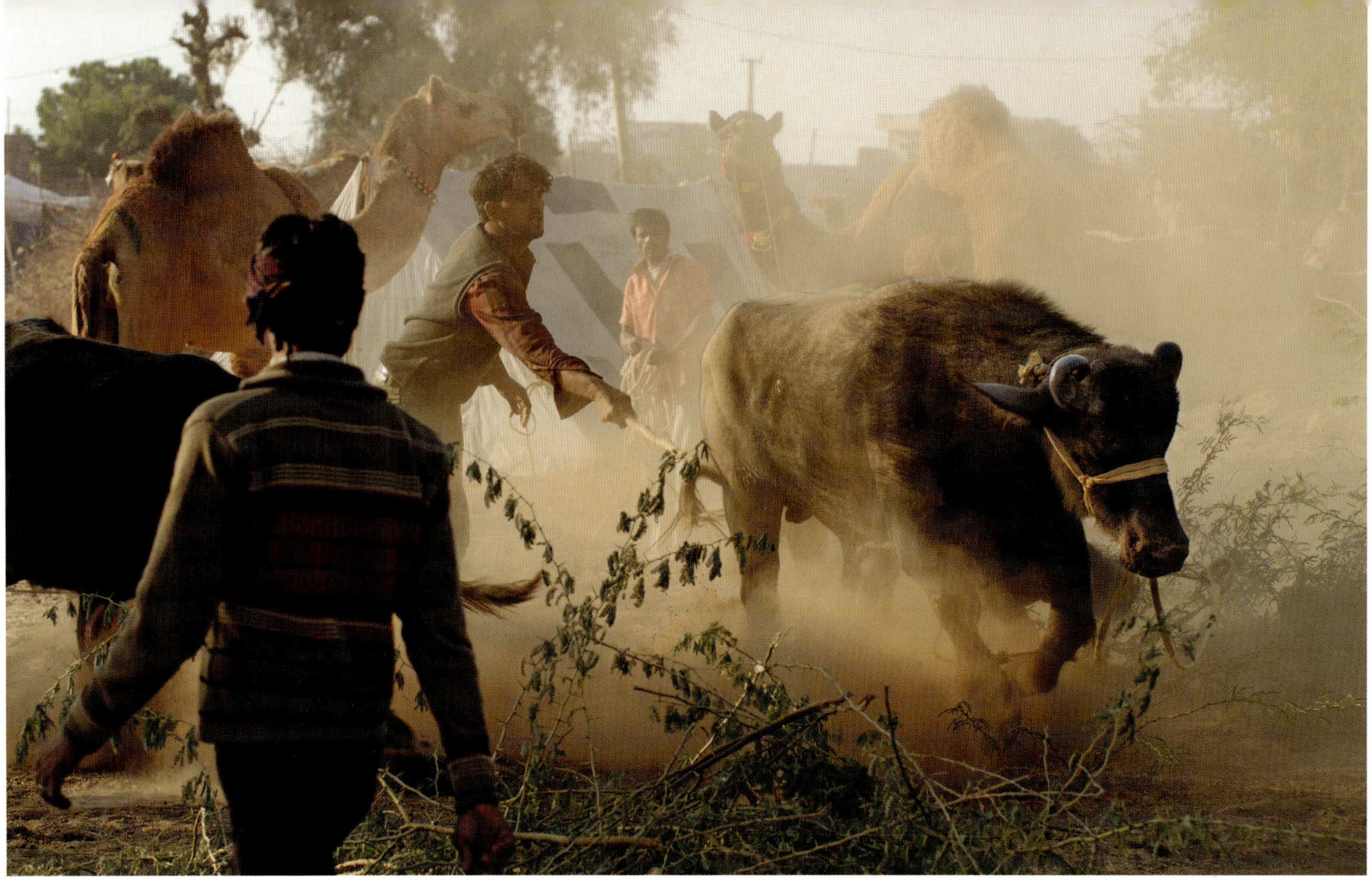

A bolting water buffalo at the camel fair can only be caught and restrained by the combined efforts of the herdsmen.

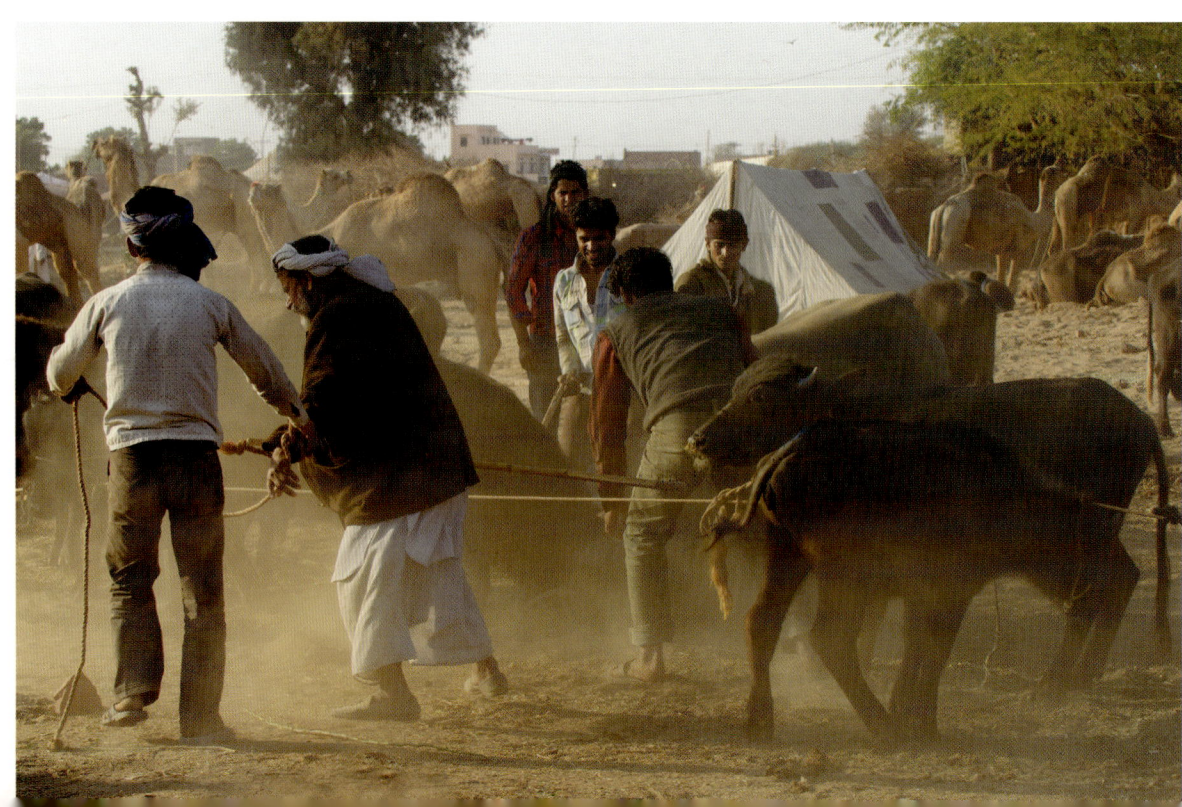

The people of the Thar Desert often live in simple conditions in mud huts. Several generations live together in a small space, sharing their quarters with their livestock.

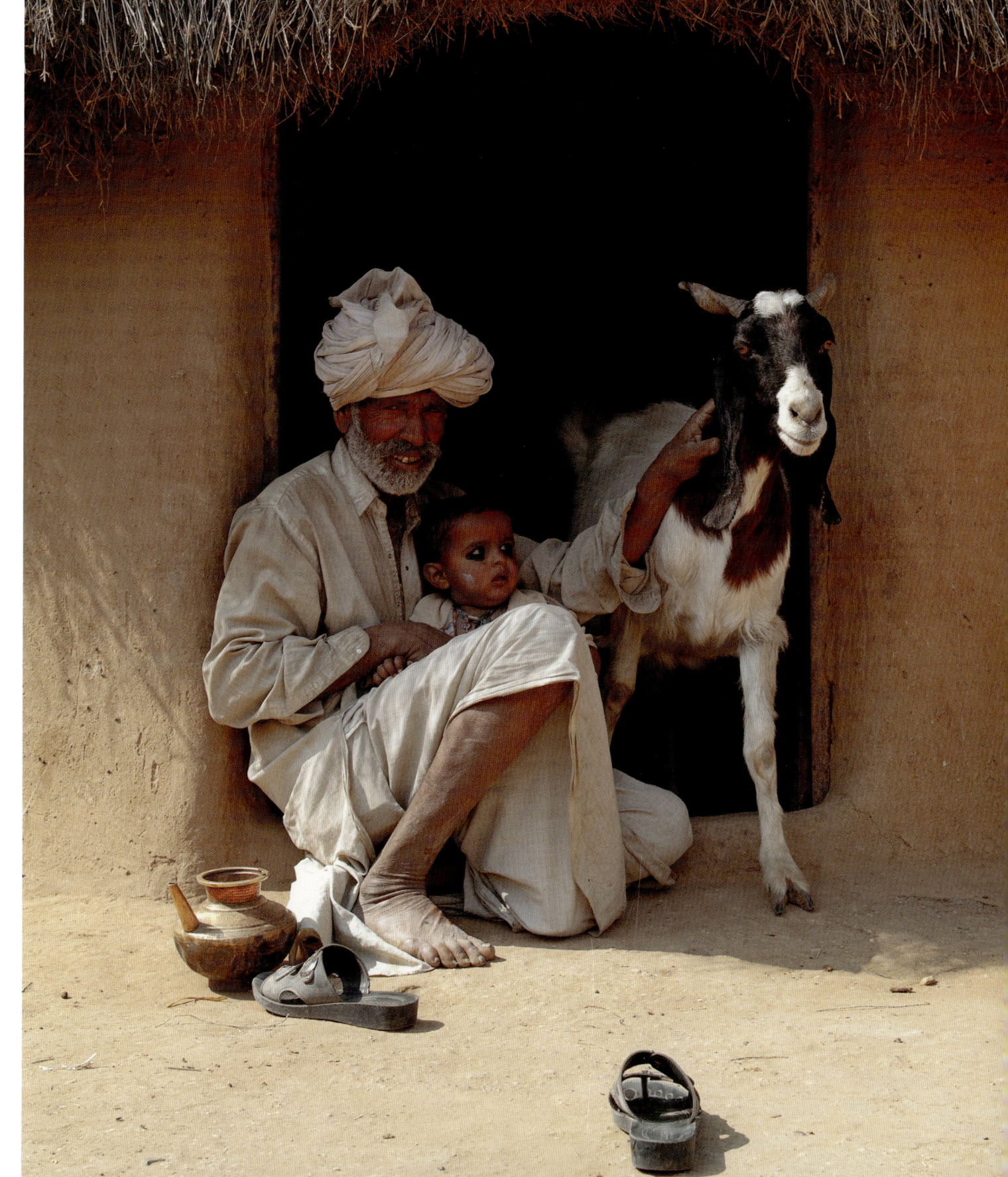

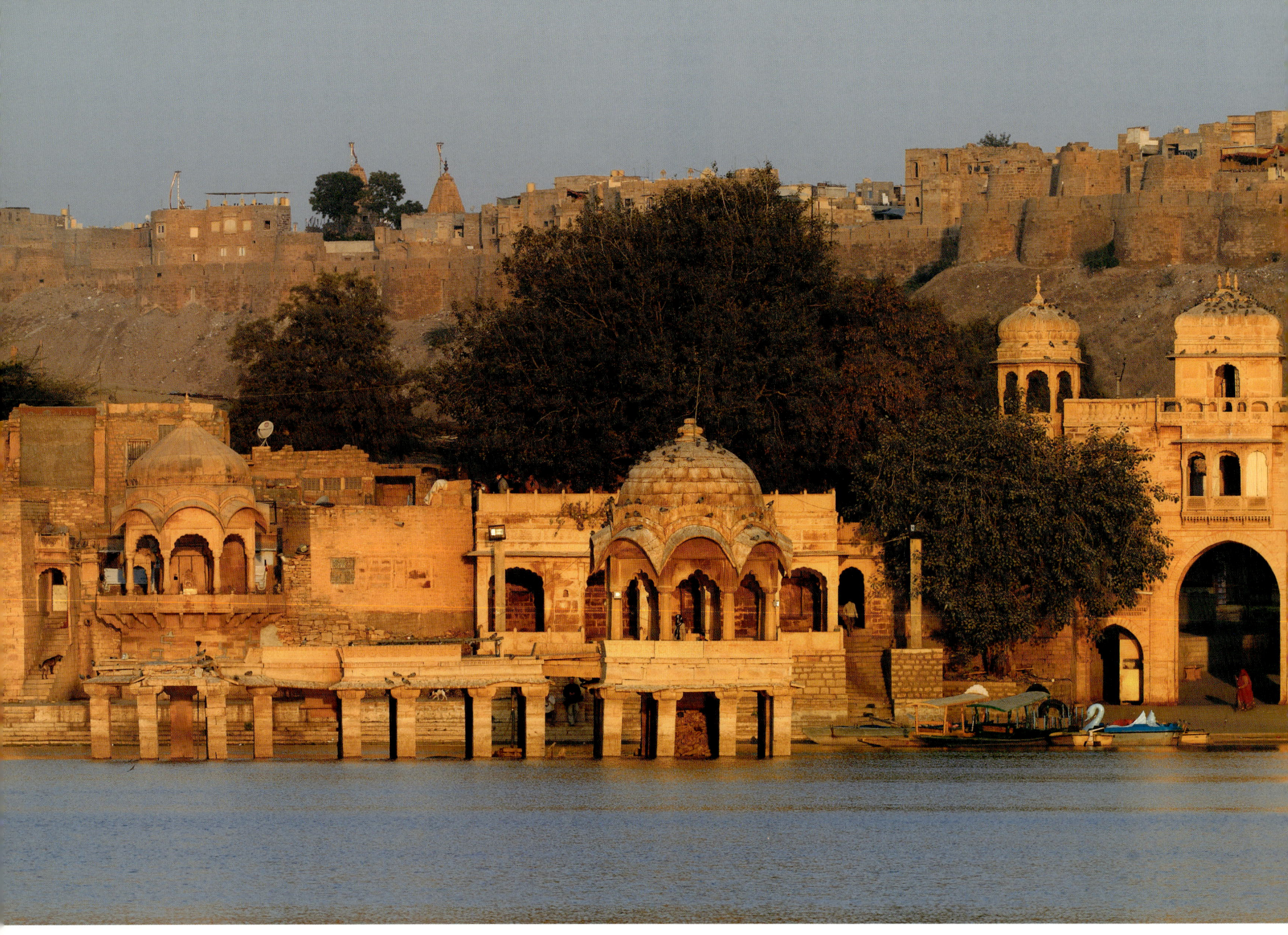

Jaisalmer looks like a movie set for a tale from *One Thousand and One Nights*. From the glowing sandstone of its buildings it has acquired the nickname of the 'Golden City'. It was founded in the year 1156.

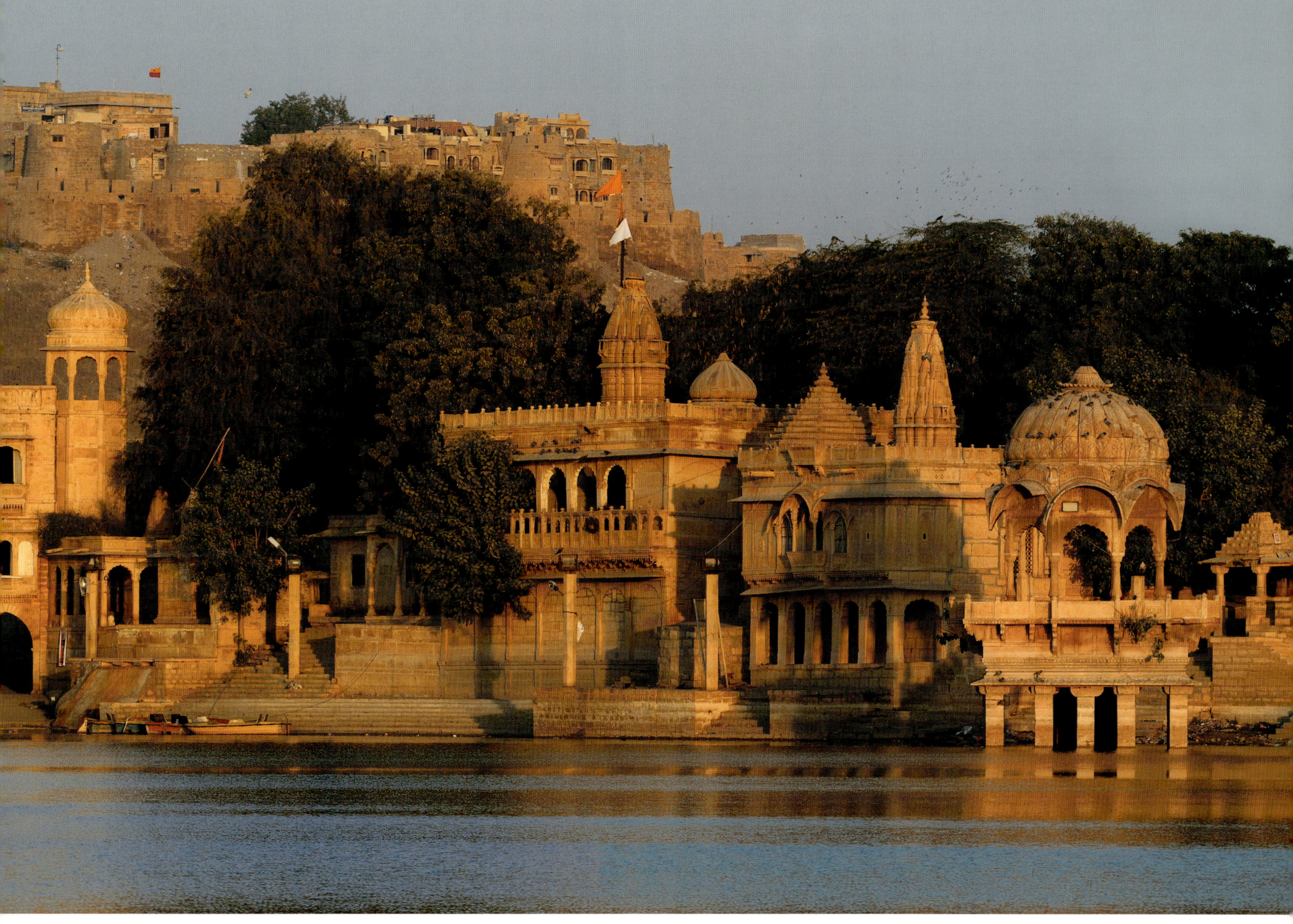

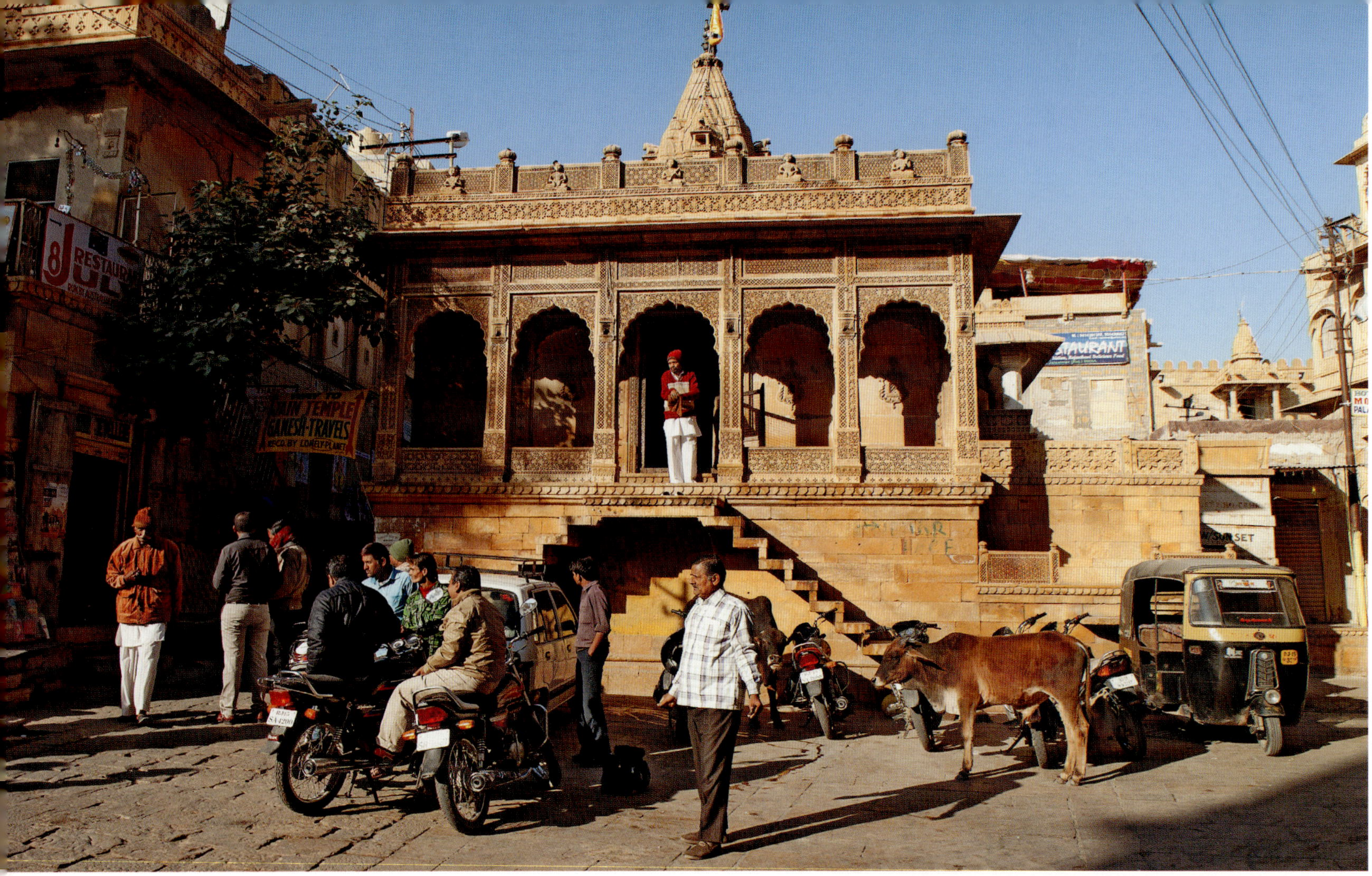

The well-preserved historic city centre of Jaisalmer is still inhabited. Within the ramparts, there are elaborately ornamented temples, while typical Indian street vendors set up their stalls outside.

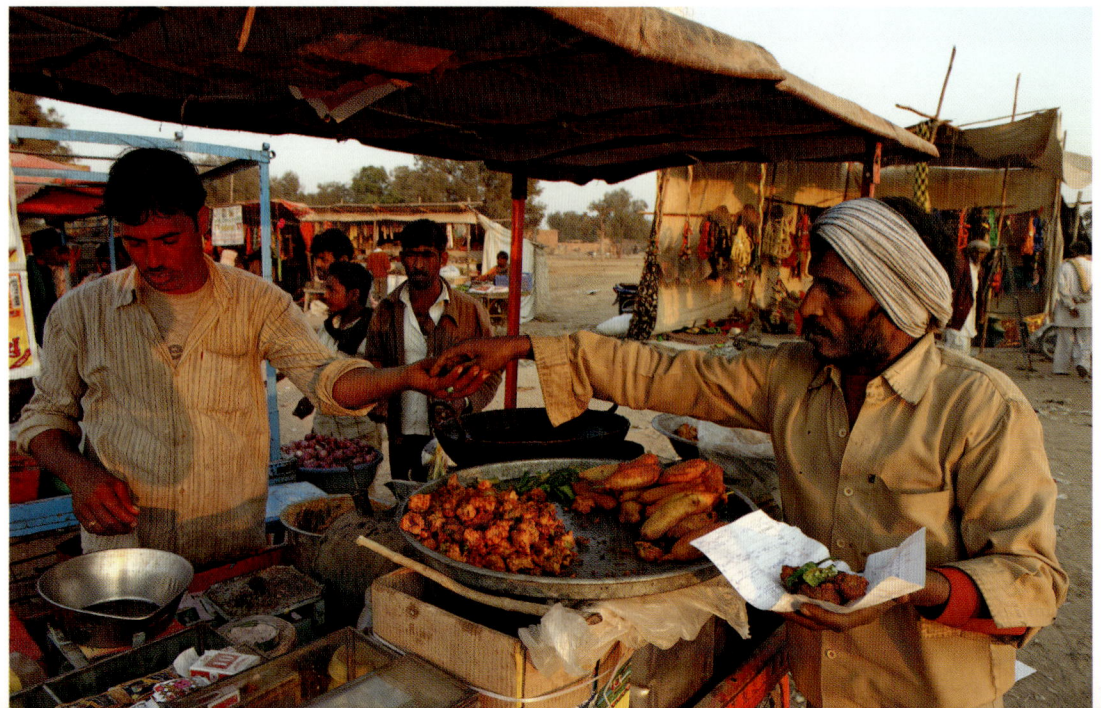

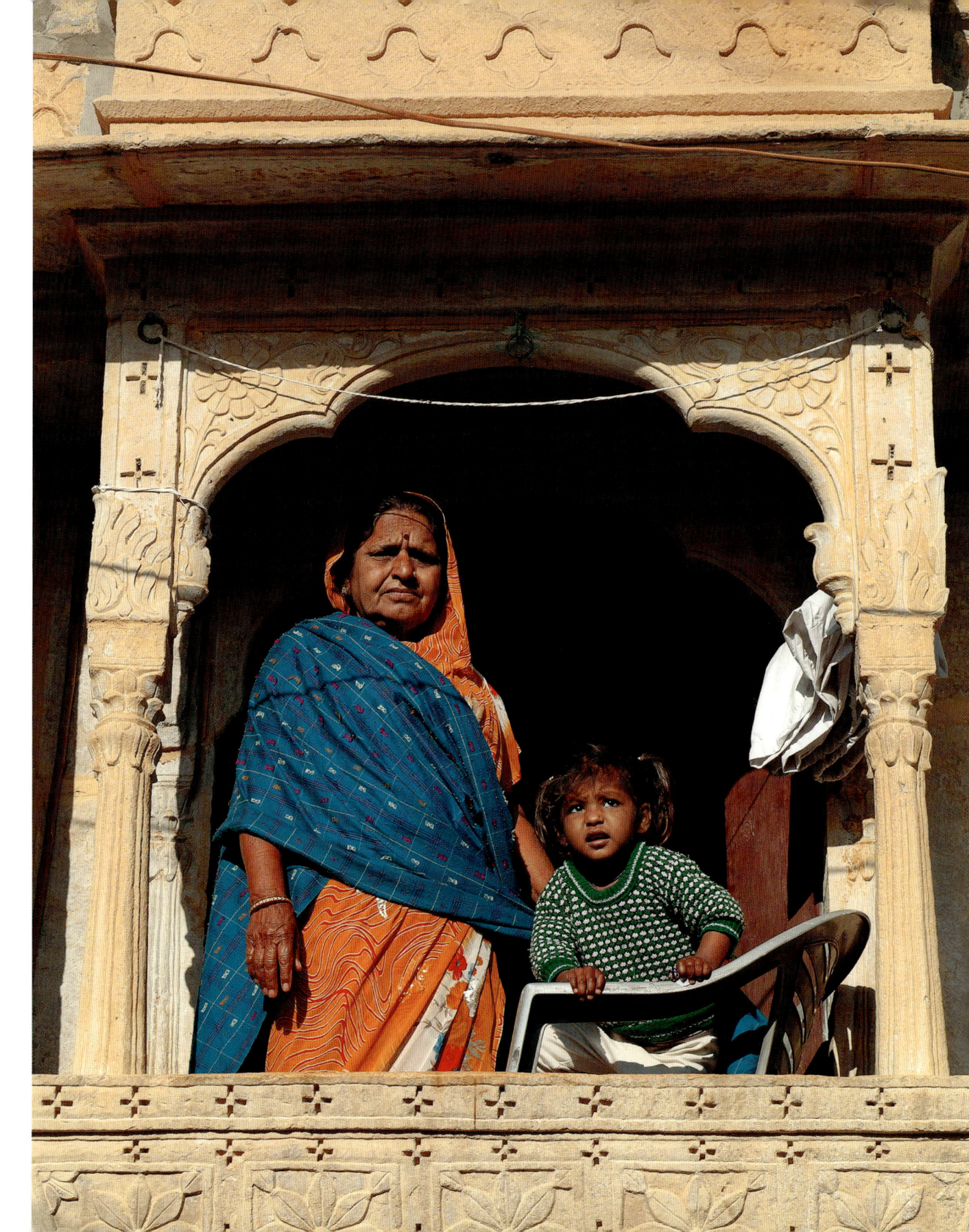

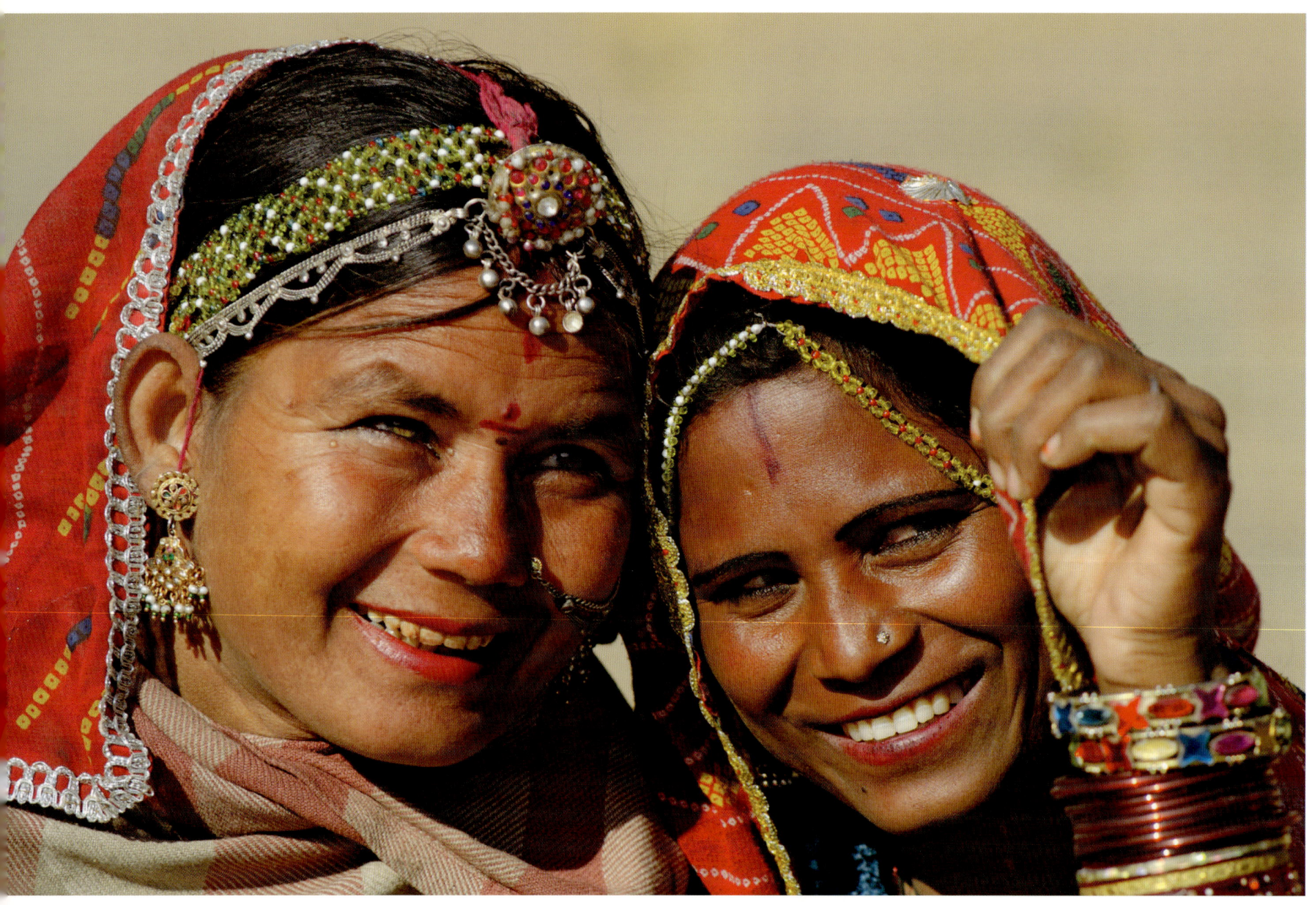

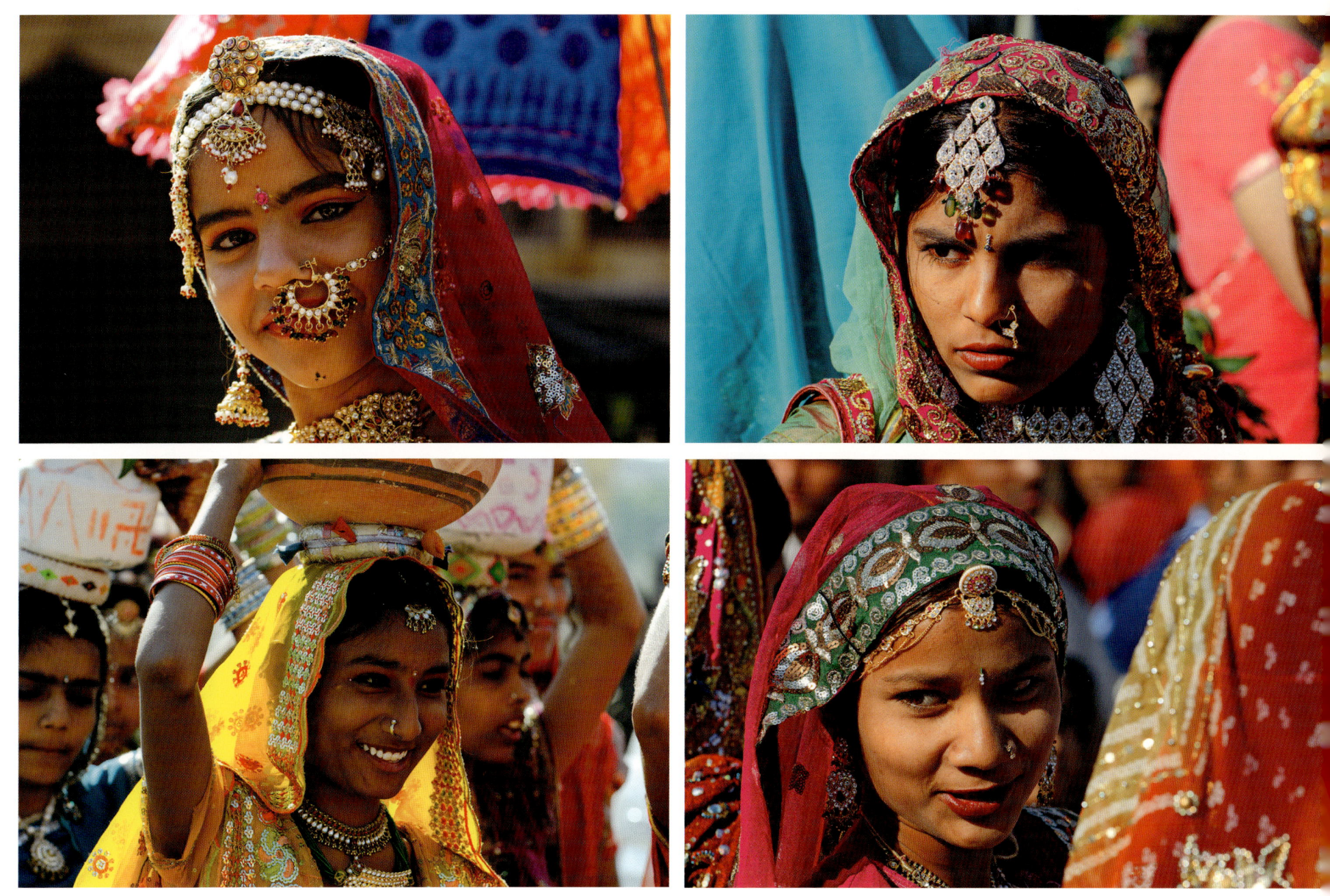

In Jaisalmer, many girls and women still wear traditional costumes and elaborate jewellery, most of which is painstakingly handcrafted.

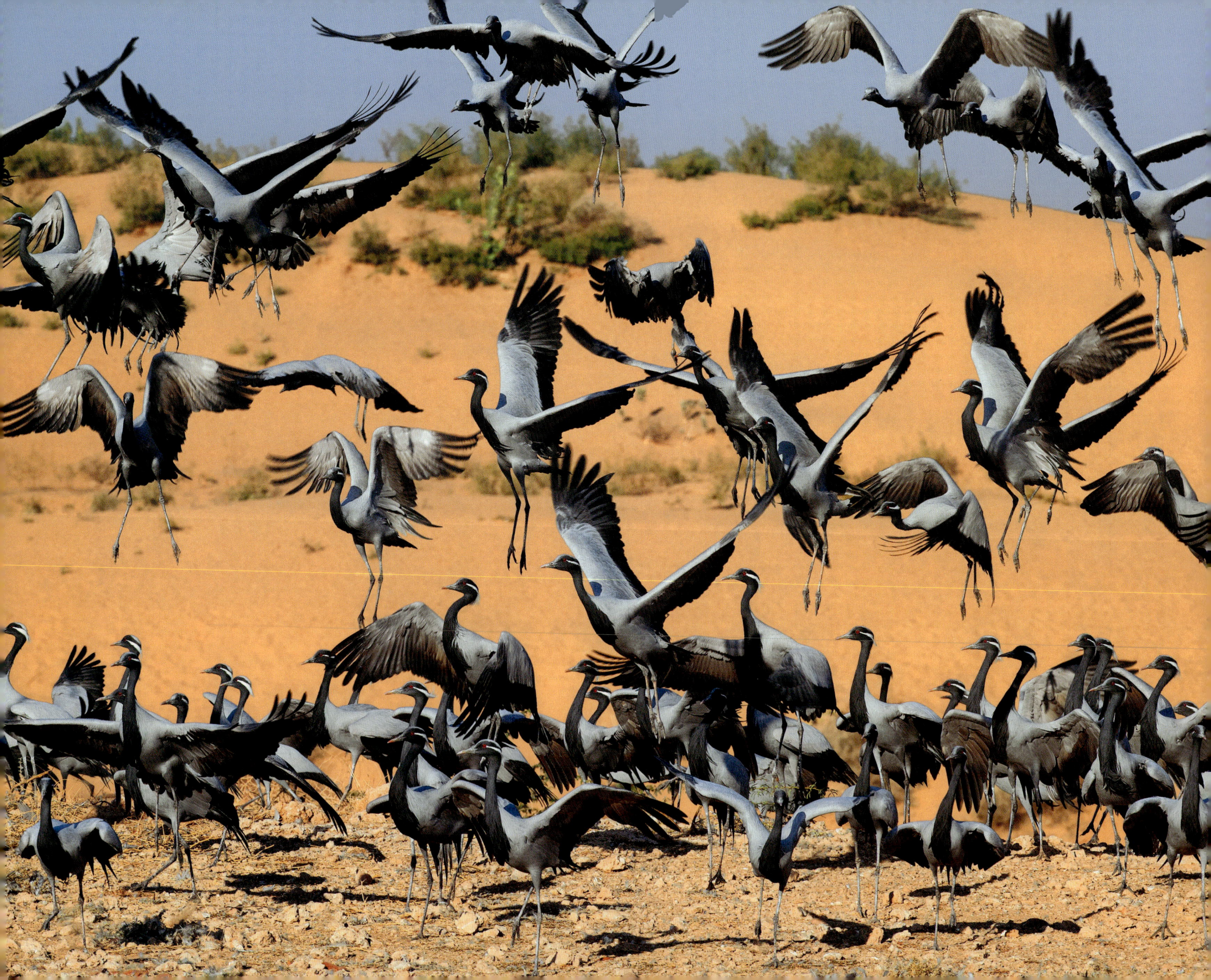

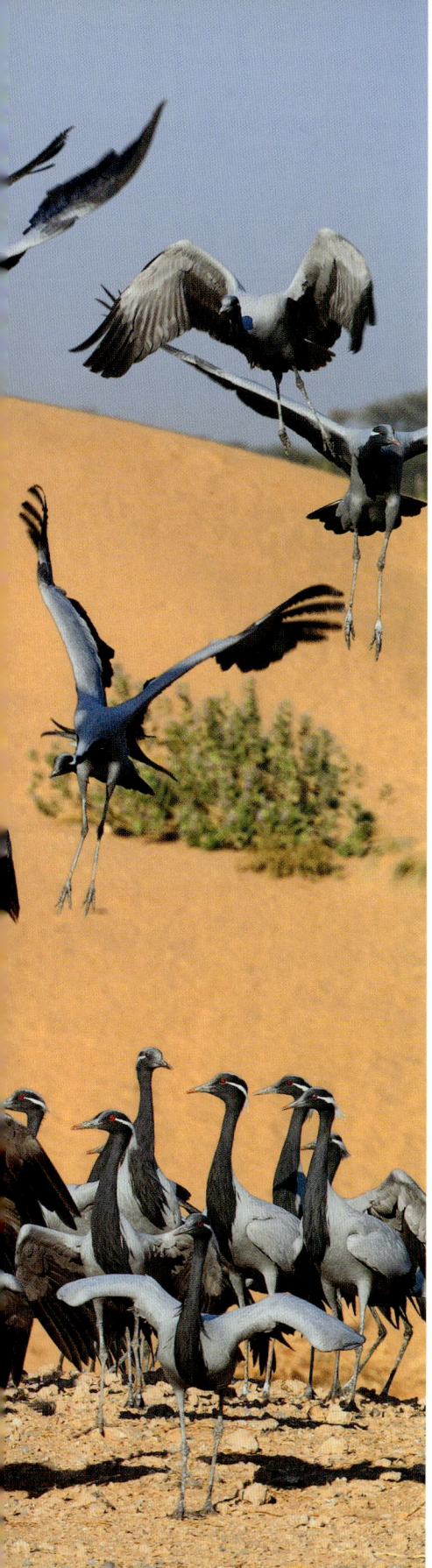

Birds of Good Fortune

India

In Rajasthan a unique spectacle takes place every year, as thousands of wintering demoiselle cranes visit a small village on the edge of the Thar Desert.

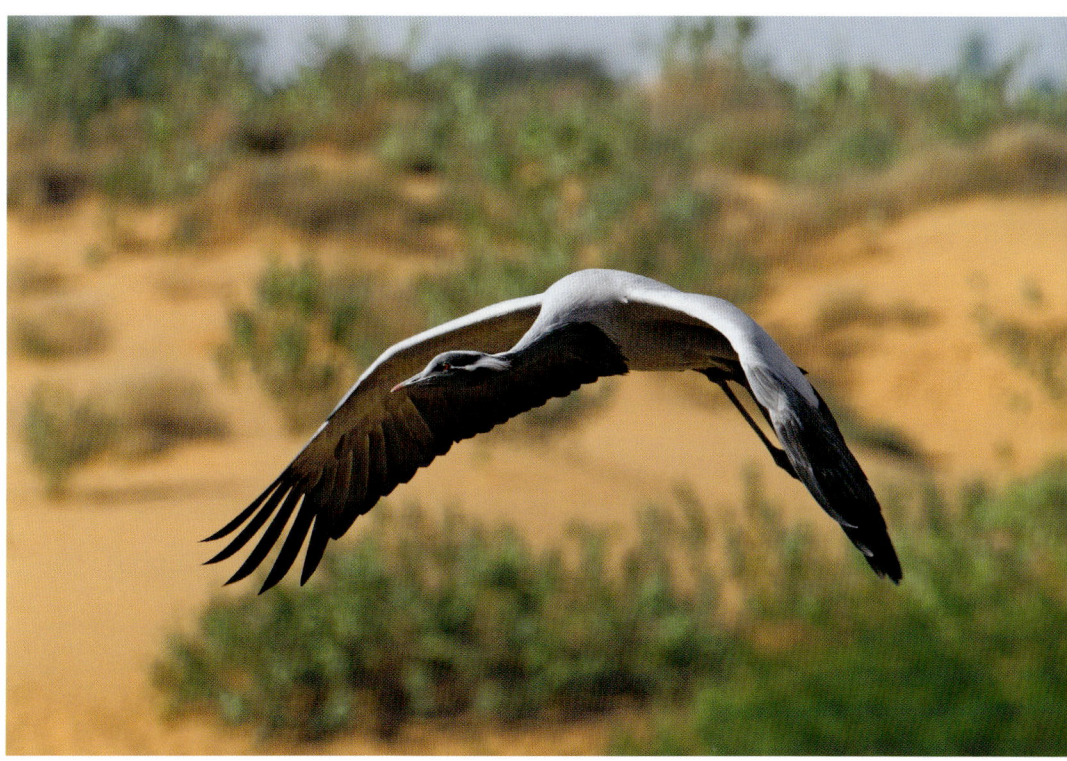

The Spectacle of the Cranes

With ear-piercing calls a peacock greets the rising sun. Slowly an old man wearing an orange-coloured turban shuffles across the village square and begins to scatter grain – food for the birds of good fortune. Pigeons start pecking at the first seeds and the old man allows this – though the food is actually meant for the cranes. There is as yet no hint of how many birds will soon come crowding in. But no sooner has the old man finished scattering the grain from his bags, than the first squadrons of demoiselle cranes descend on the village. With loud trumpeting calls they land at the feeding ground. More and more birds arrive: first dozens, then hundreds and eventually thousands. The large birds greedily peck up the scattered grain. All the while more battalions arrive. The hungry birds cannot all feed at once – there are just too many of them. And when a flock has eaten its fill and makes room for the next, the masses of birds seem to explode into the air. This visual treat is accompanied by the background noise of swooshing wings and the constant trumpeting of the cranes.

This magnificent spectacle has only been a feature of this village for the last few years. Demoiselle cranes have always come to this part of India to escape the cold winters of their central Asian breeding grounds, but there were traditionally just a few dozen birds around the village of Khichan. As the villagers began feeding them, more and more birds were attracted and their numbers gradually increased here. Today, between 12,000 and 15,000 demoiselle cranes congregate around the village every year. The result, of course, is that considerably more grain is required to feed them. Wealthy members of the Jain religion provide the financial means to buy the birdseed. The core area for Jainism is in India, but a few devotees have become affluent as businessmen overseas. Their willingness to donate to charity and do good deeds is founded in their religion.

Jains are peace-loving people who are forbidden to harm or kill any form of life. Many of them believe in the traditional idea that all creatures on earth are part of one large family. Furthermore, in India cranes symbolize happiness in marriage – presumably because the birds often form pair bonds that last for life. Cranes appear in folk songs and poems, they are venerated and are considered to be birds that bring good luck. This religiously motivated care for the demoiselle cranes has developed into one of the most unusual relationships between humans and wild animals in the whole of India.

The villagers have even fenced in the feeding ground, so that dogs or boisterous children do not unnecessarily disturb the birds. As the feeding ground is open at the top, the cranes can come and go as they please. Occasionally the odd bird has been caught up in the power supply lines and died. Some power lines have therefore been insulated or even removed to reduce the risk of injury, with more to follow.

One very practical aspect of the feeding is that it enables the villagers to control the birds: if they eat their fill at the feeding ground, they will not go near the farmers' crops. Conflicts with the locals are largely avoided. After their morning feed the cranes roost for a few hours on the sand dunes out in the Thar Desert, and at night they retreat to a distant lake. But come morning they will be back again for their next feed, and everybody rejoices in the return of the birds of good fortune.

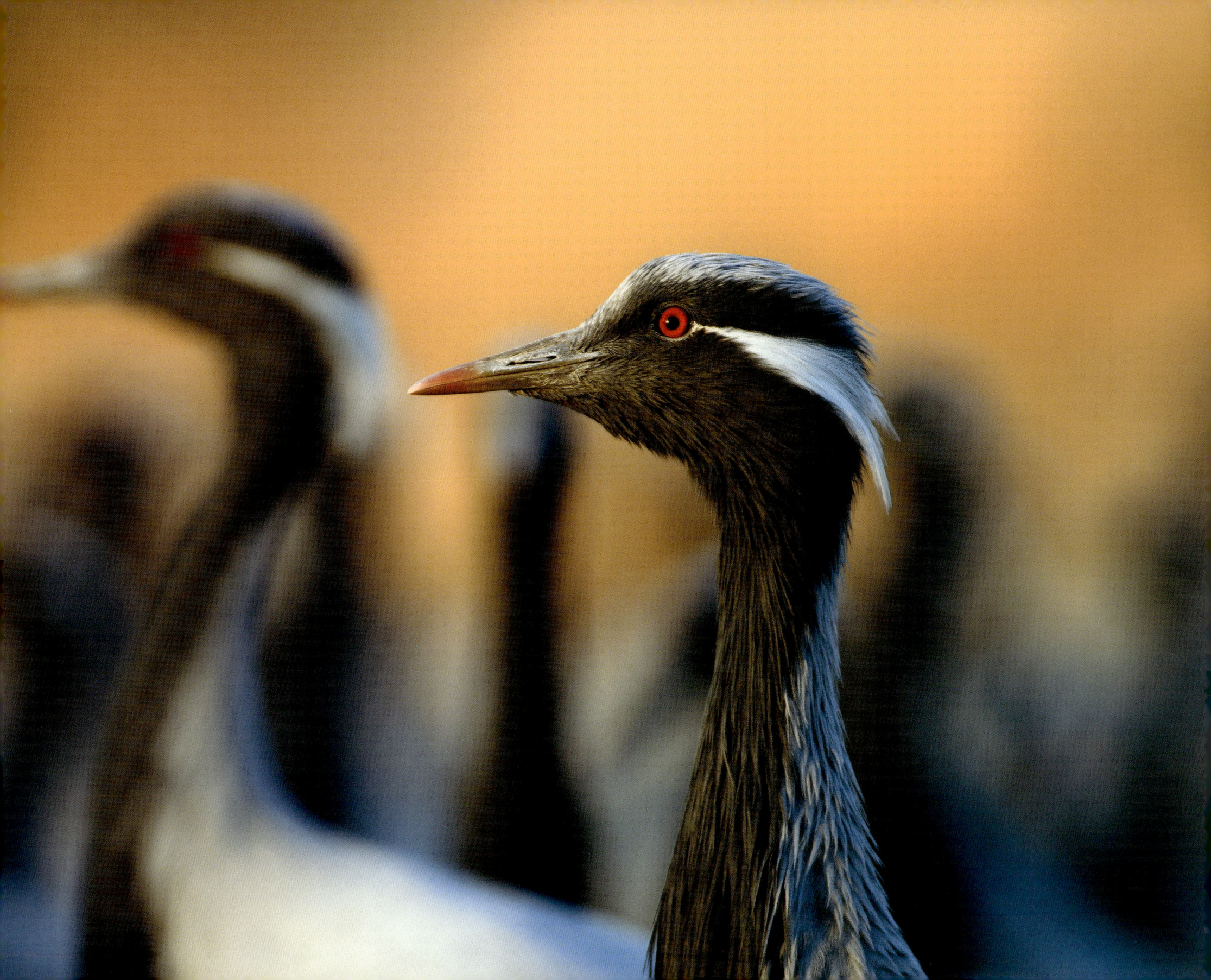

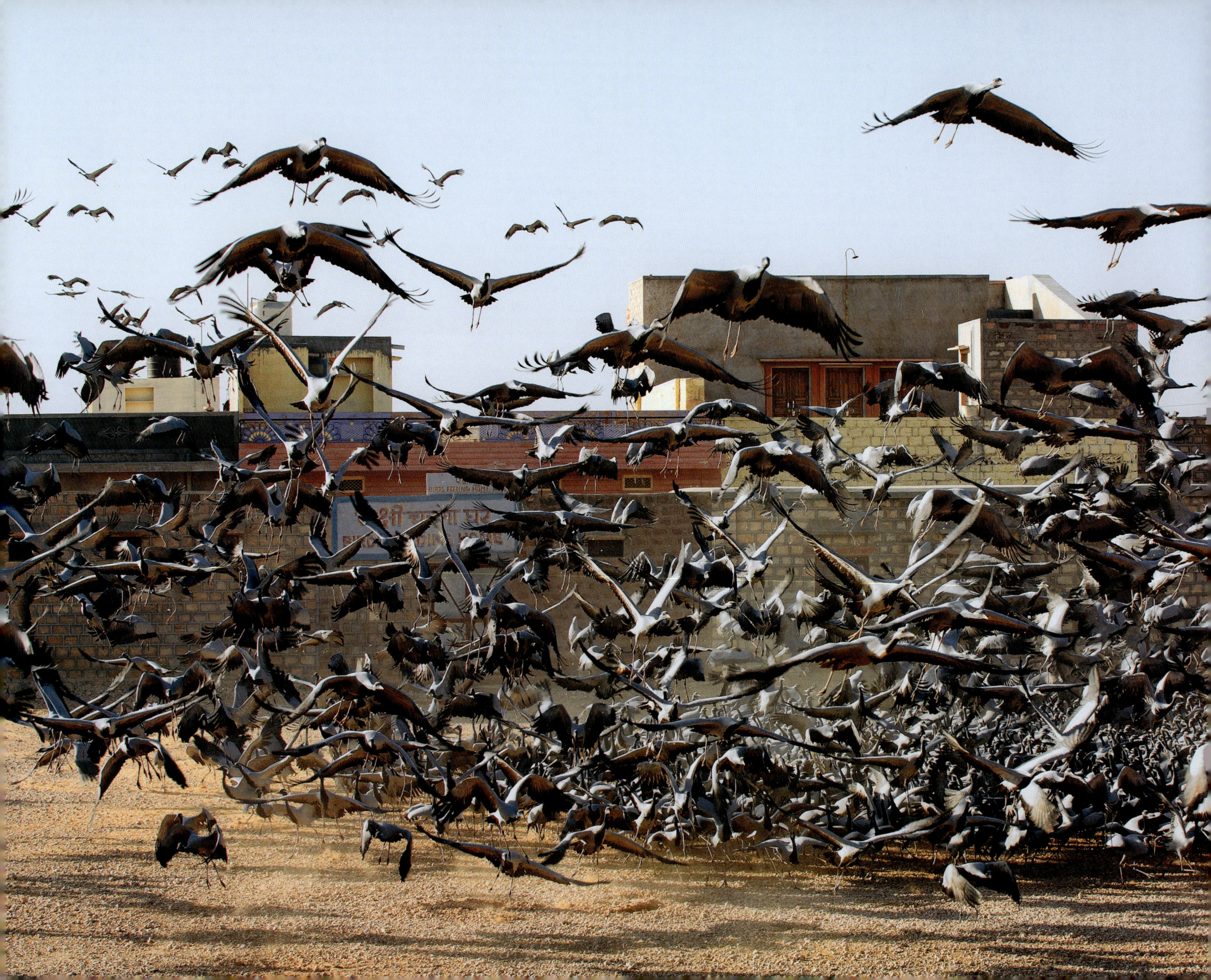

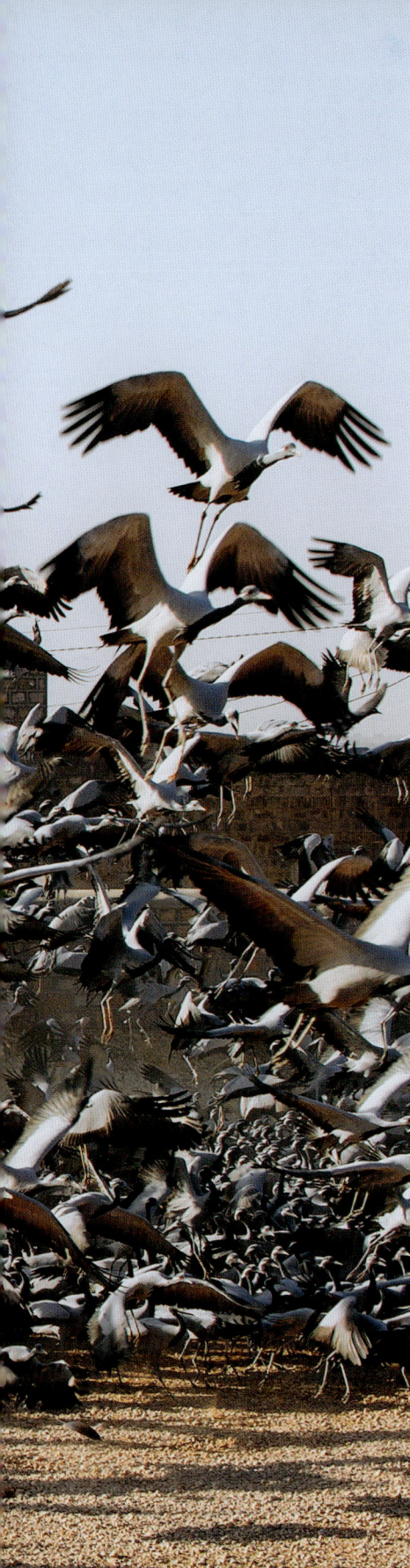

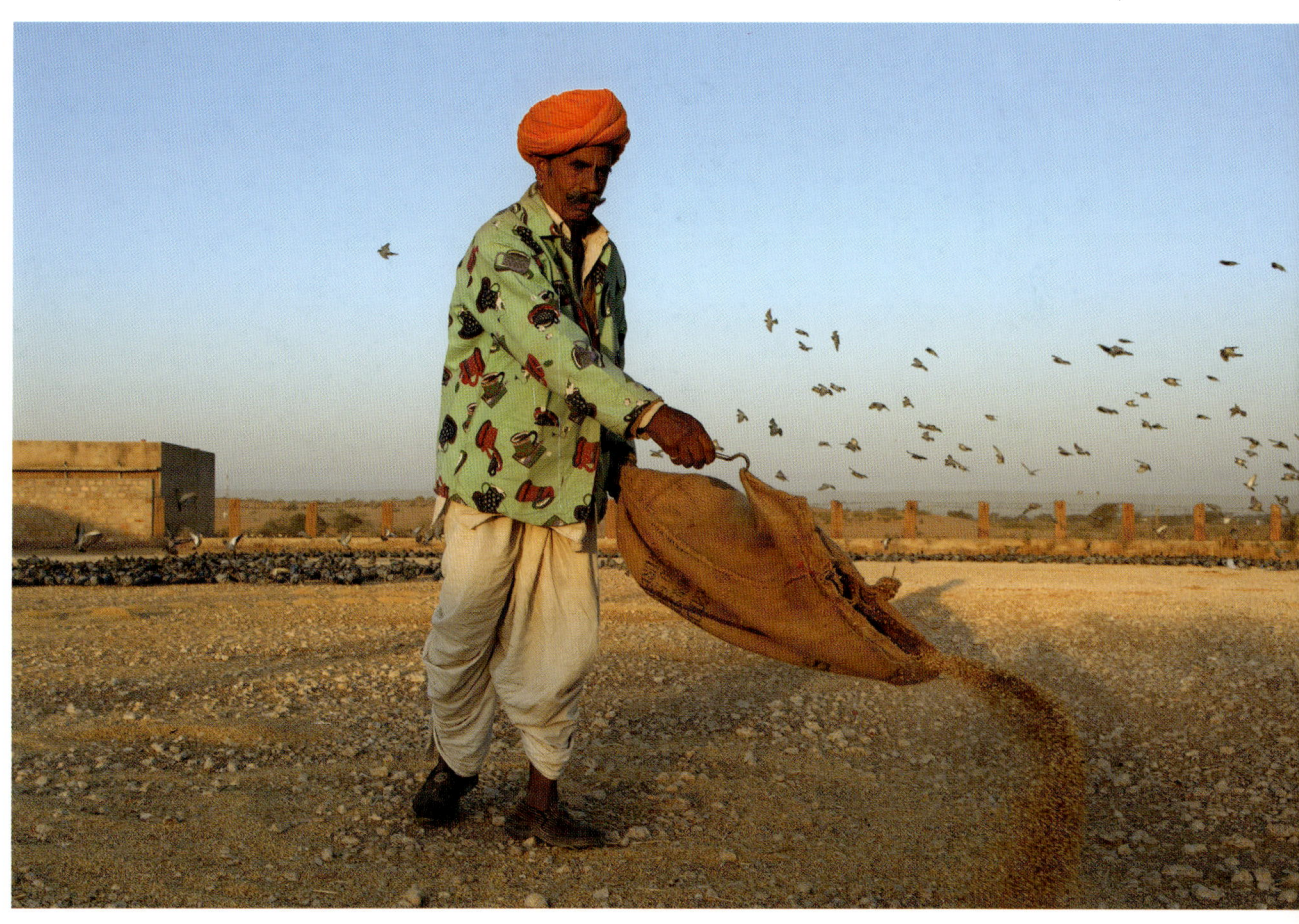

Every day the people of the village of Khichan scatter large quantities of grain to feed the birds. Demoiselle cranes from central Asia, which come to this part of India to overwinter, therefore flock in huge numbers to the village square. When they have had their fill they depart, and the next flock of cranes can take their place.

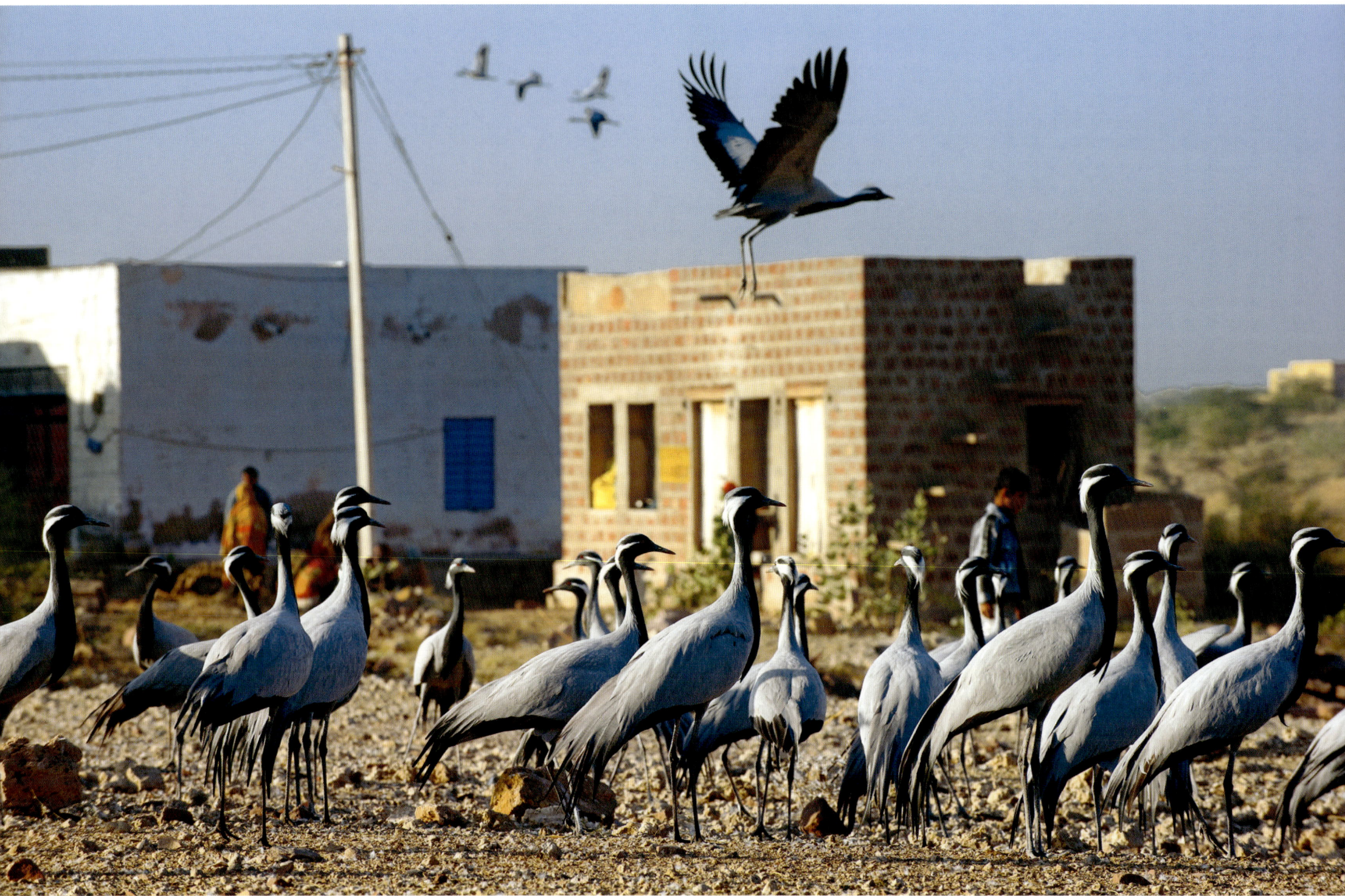

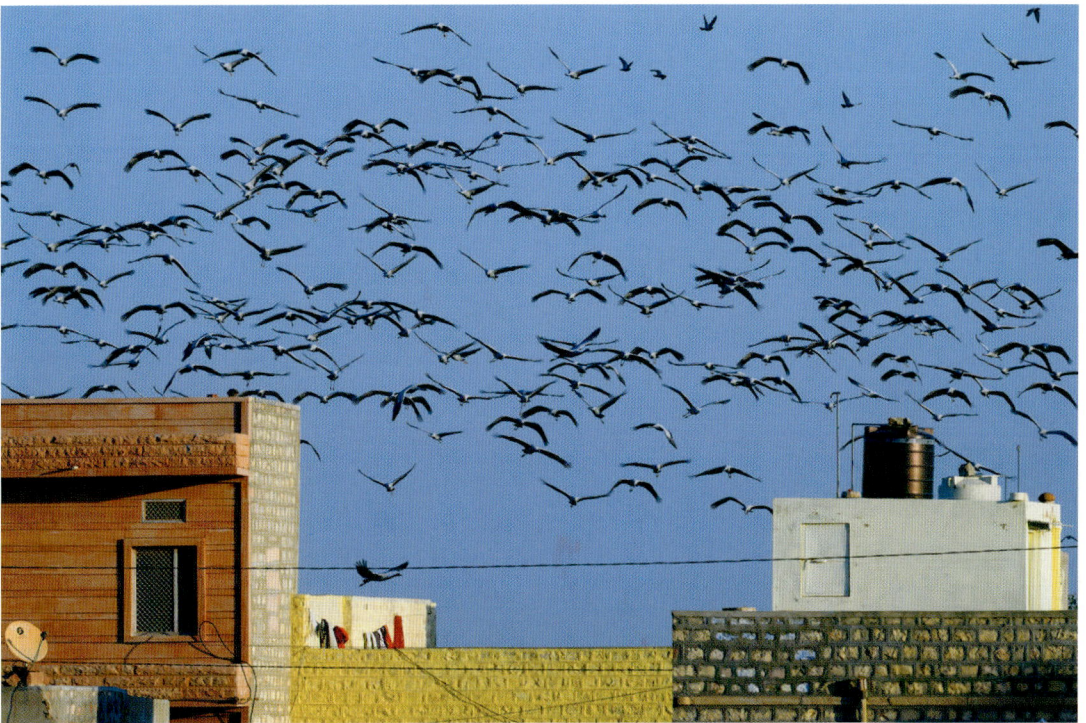

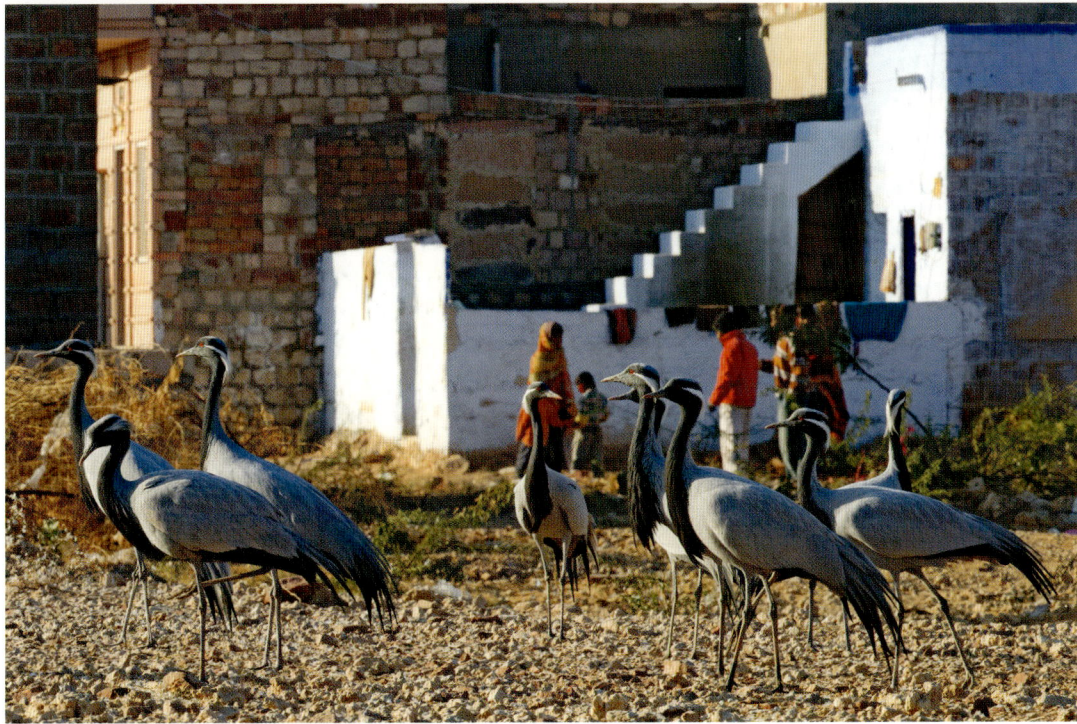

Large flocks of demoiselle cranes circle above the village and finally settle between the houses. Elsewhere, these birds are typically quite shy, but in this village they are surprisingly confiding, and show little fear in close proximity to humans.

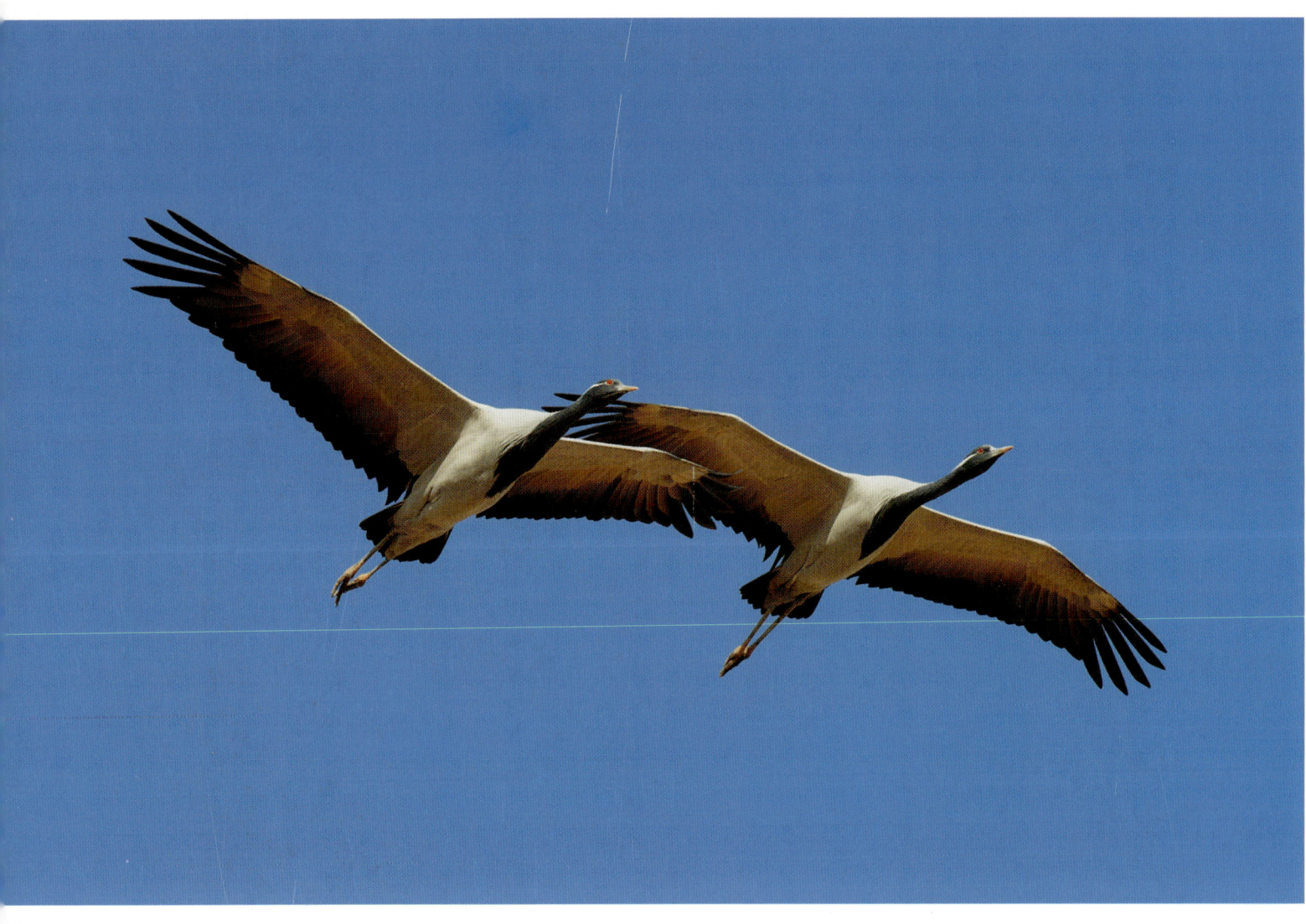

After the demoiselle cranes have fed, the morning sun, rising higher in the sky, creates thermals above the dry desert floor. The birds take advantage of the columns of warm air to get airborne, and to save energy as they soar up and away in large flocks again.

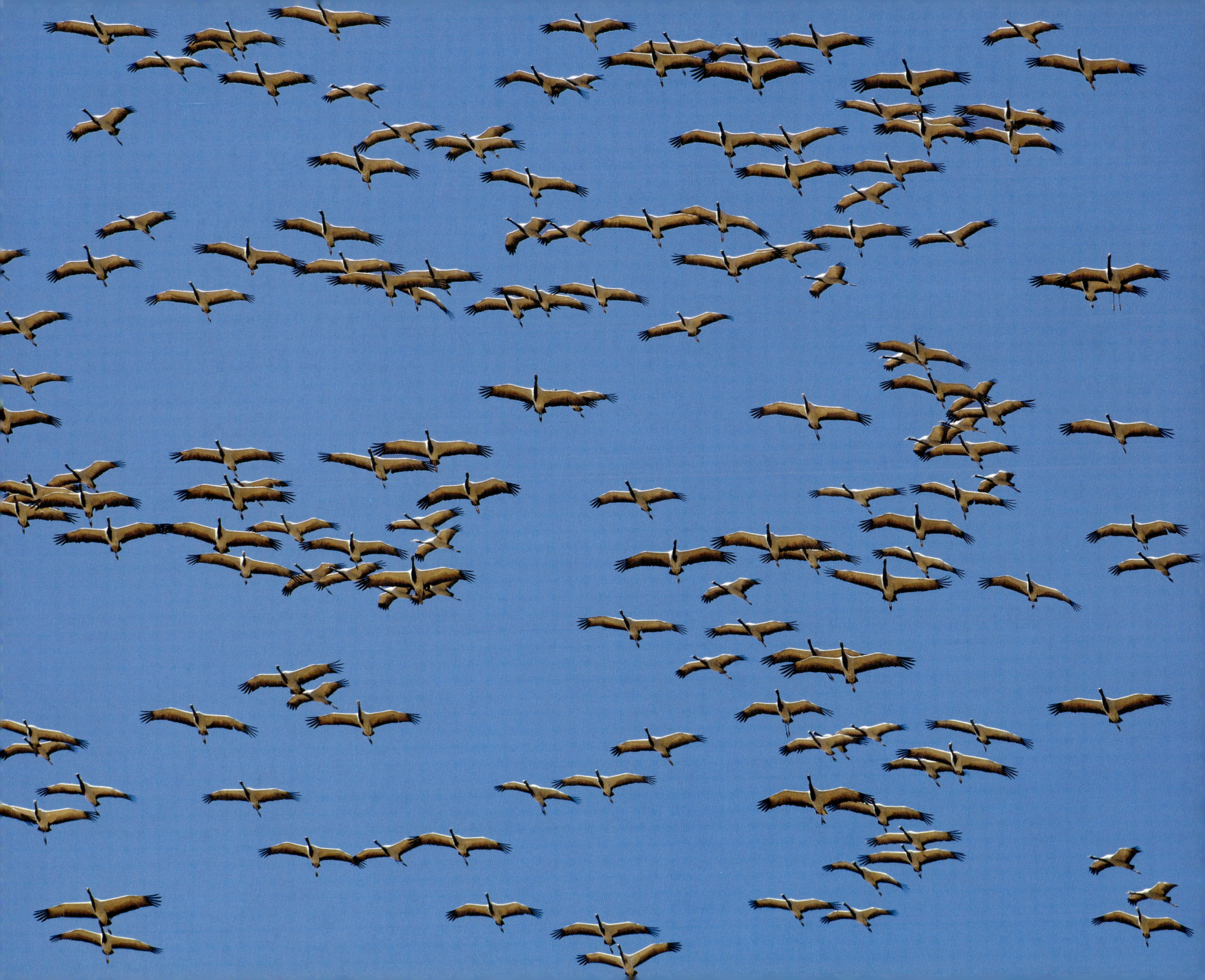

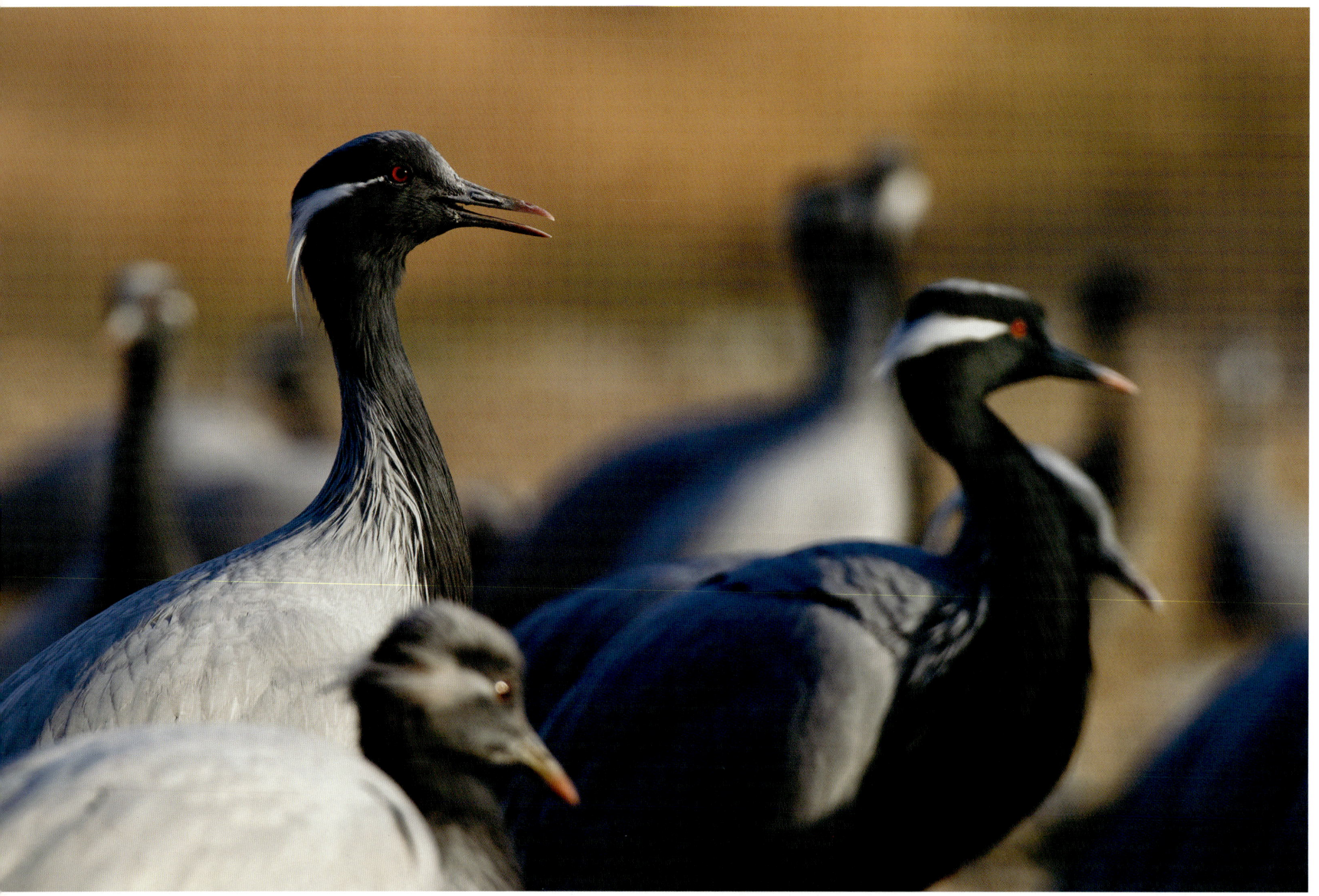

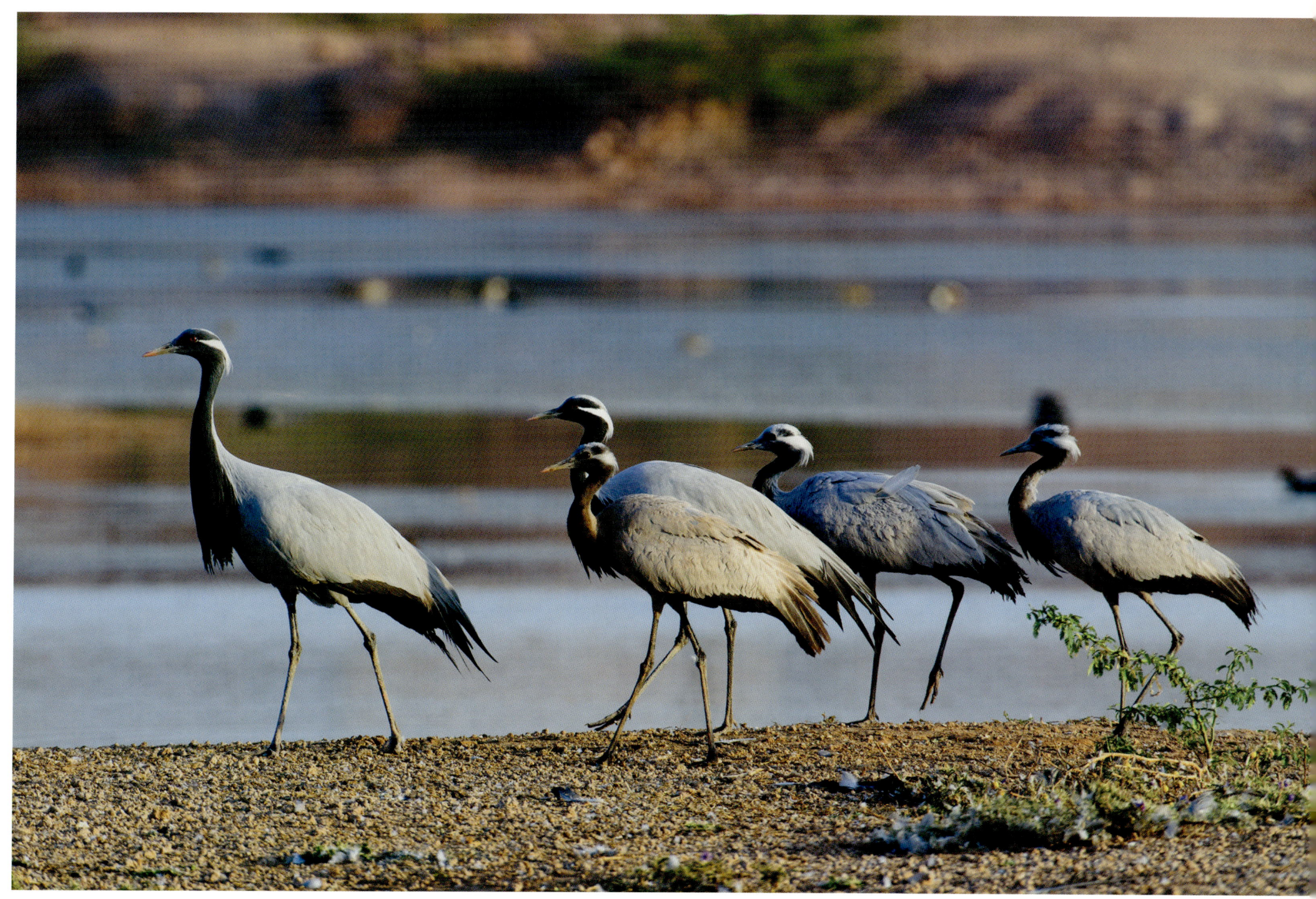

Reservoirs provide water for a little agriculture in the desert –
and here, too, the demoiselle cranes like to gather and drink.

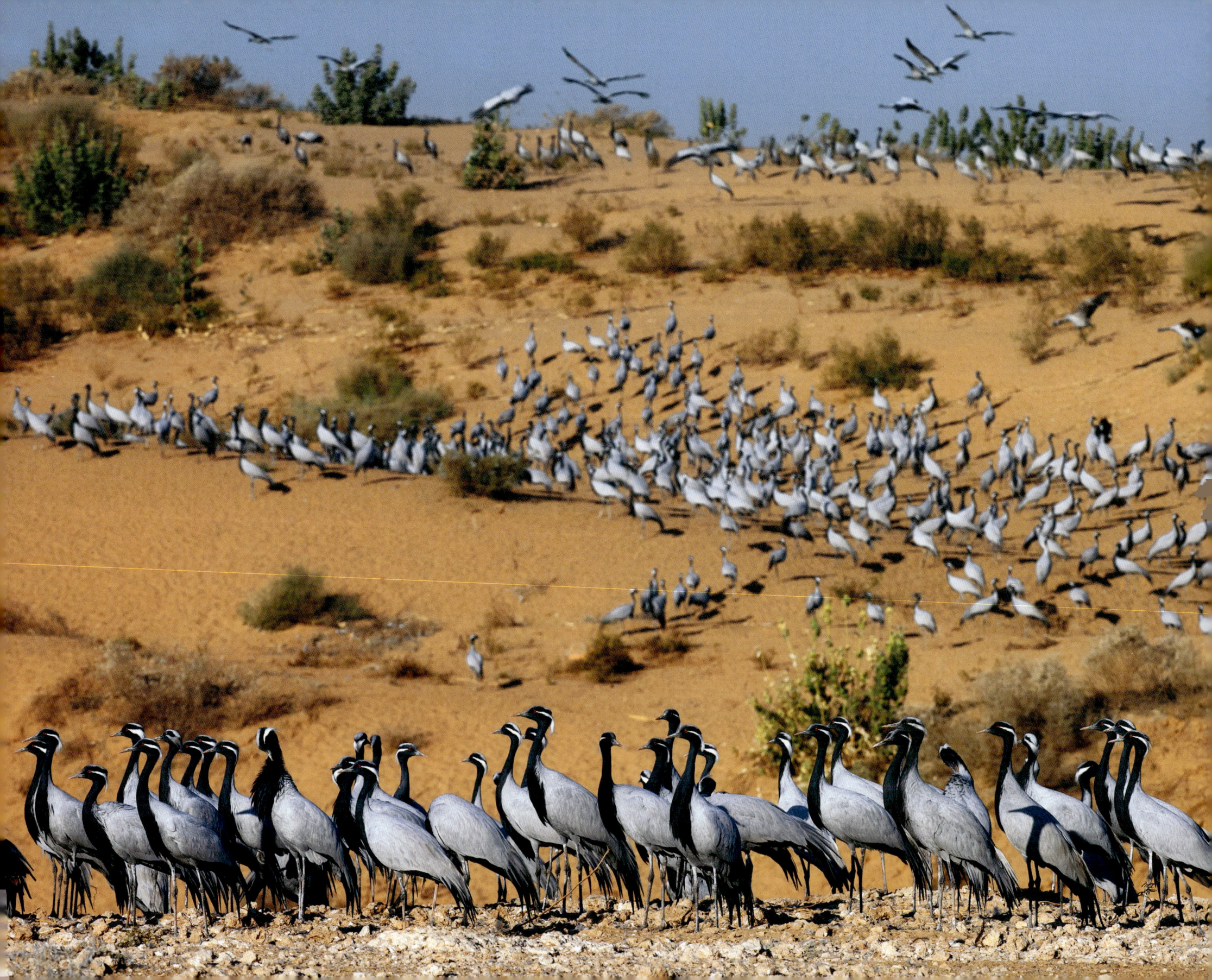

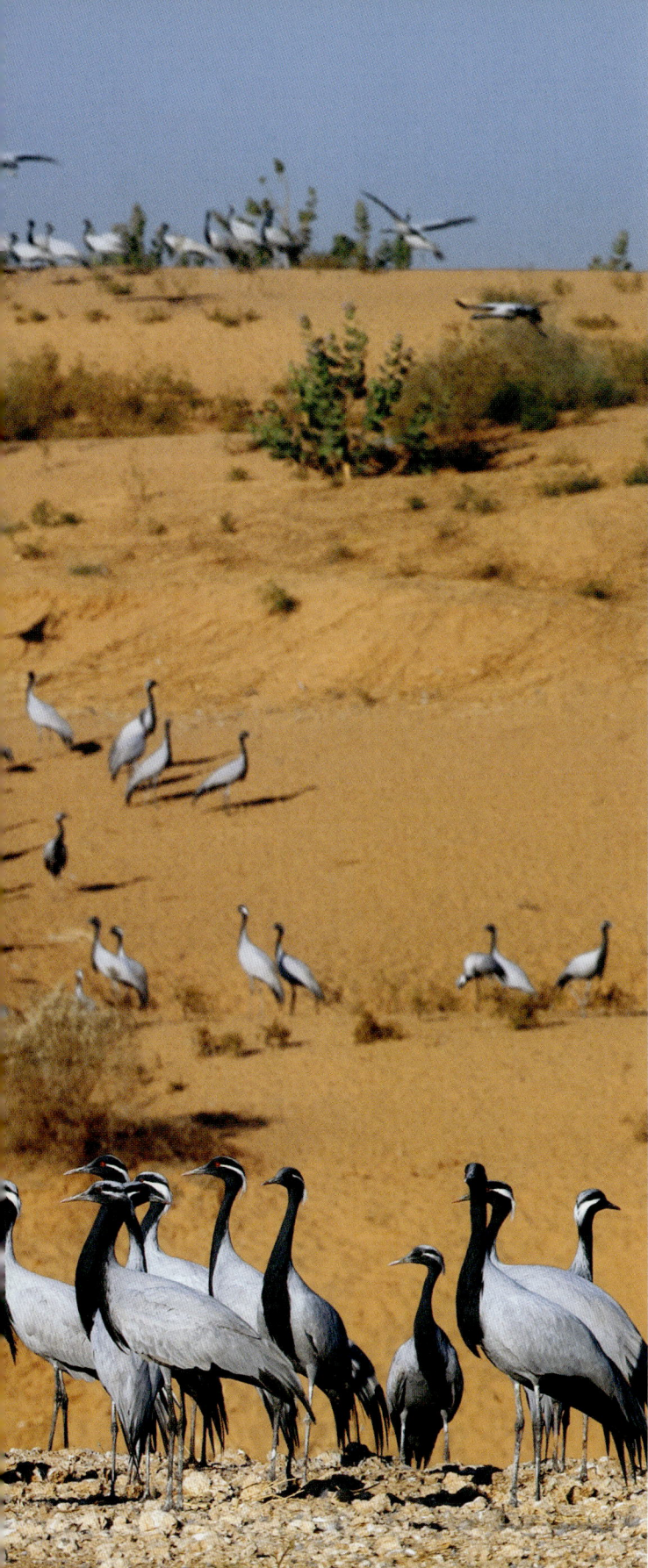

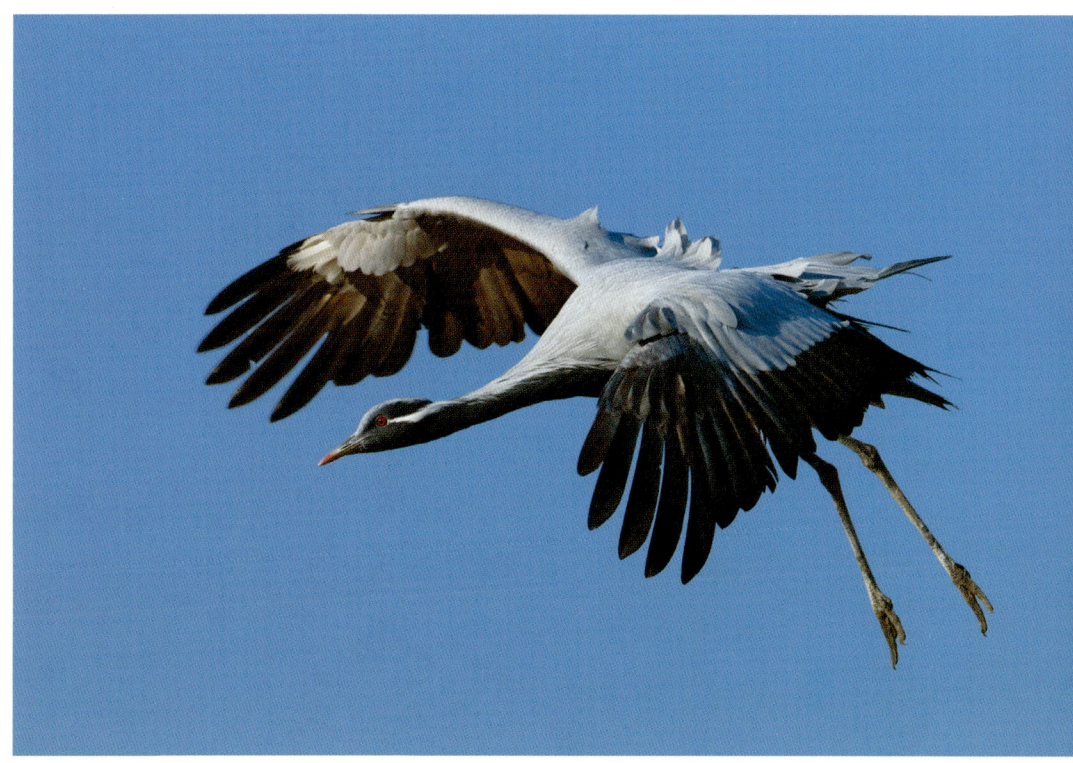

Attracted by the food that is provided, 12,000–15,000 demoiselle cranes congregate each winter in the desert around Khichan, and their numbers are rising. But every spring they return to their breeding grounds in central Asia to raise their young.

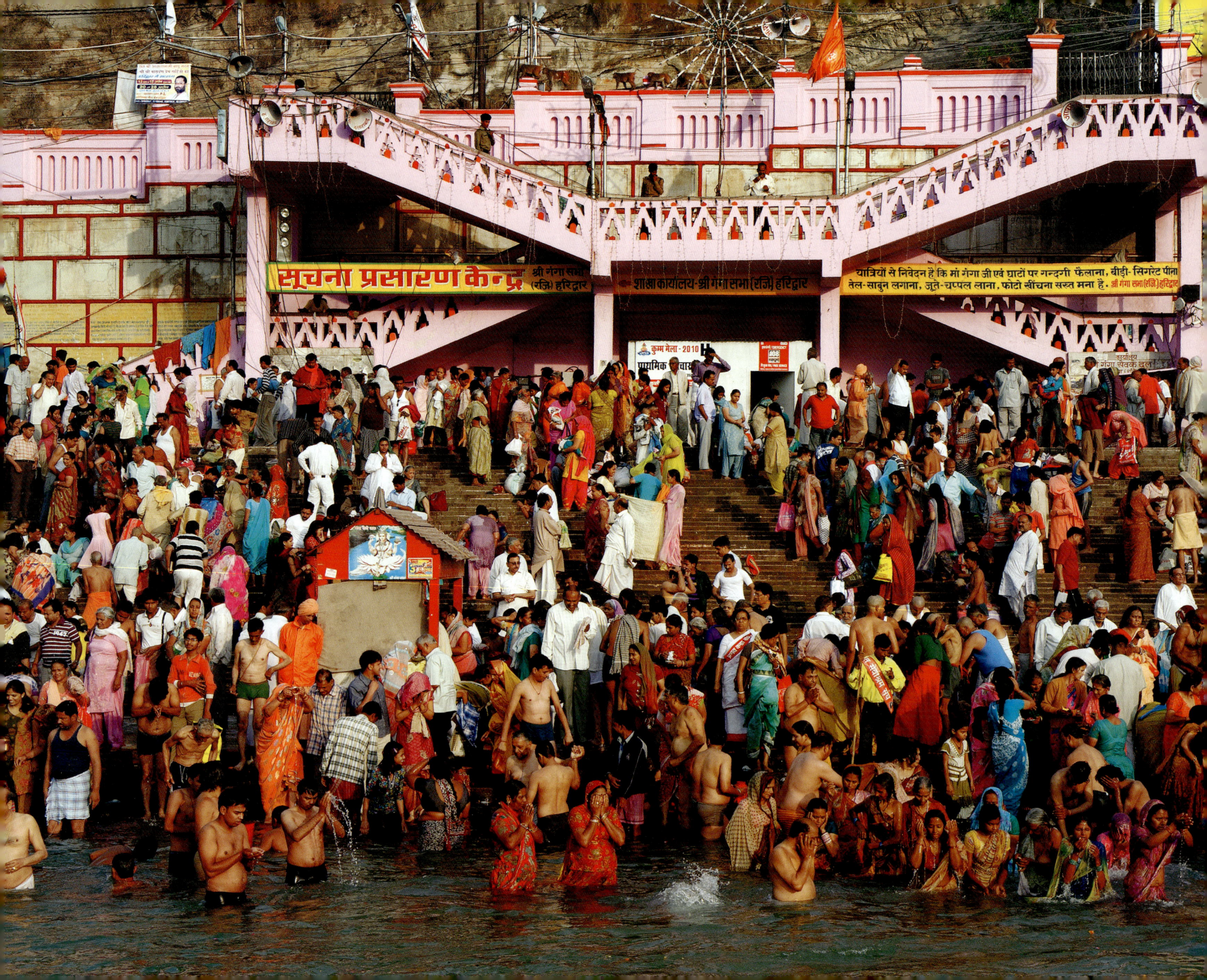

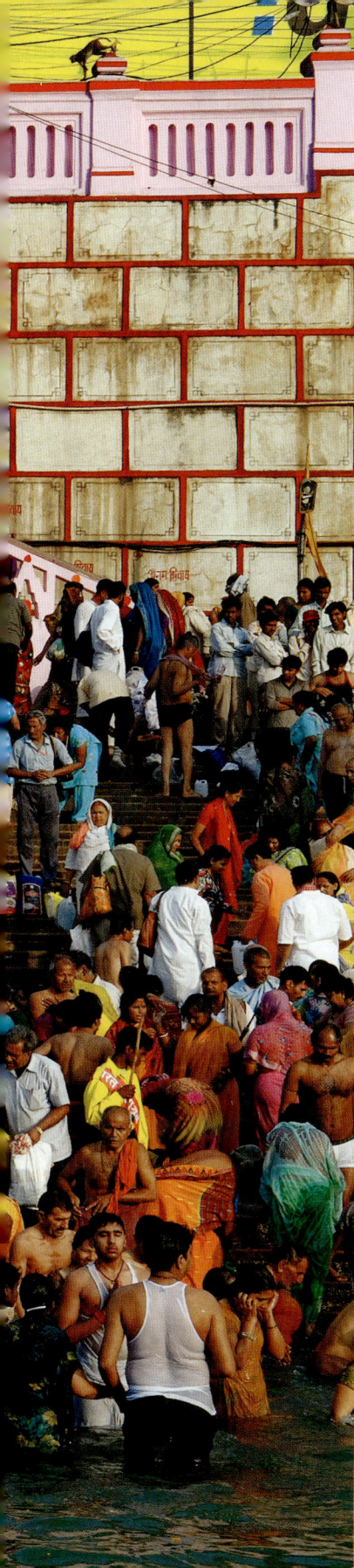

Holy Waters

India

The Ganges is a lifeline for millions of people and one of the most sacred sites in India. Countless pilgrims bathe in the river to wash away their sins.

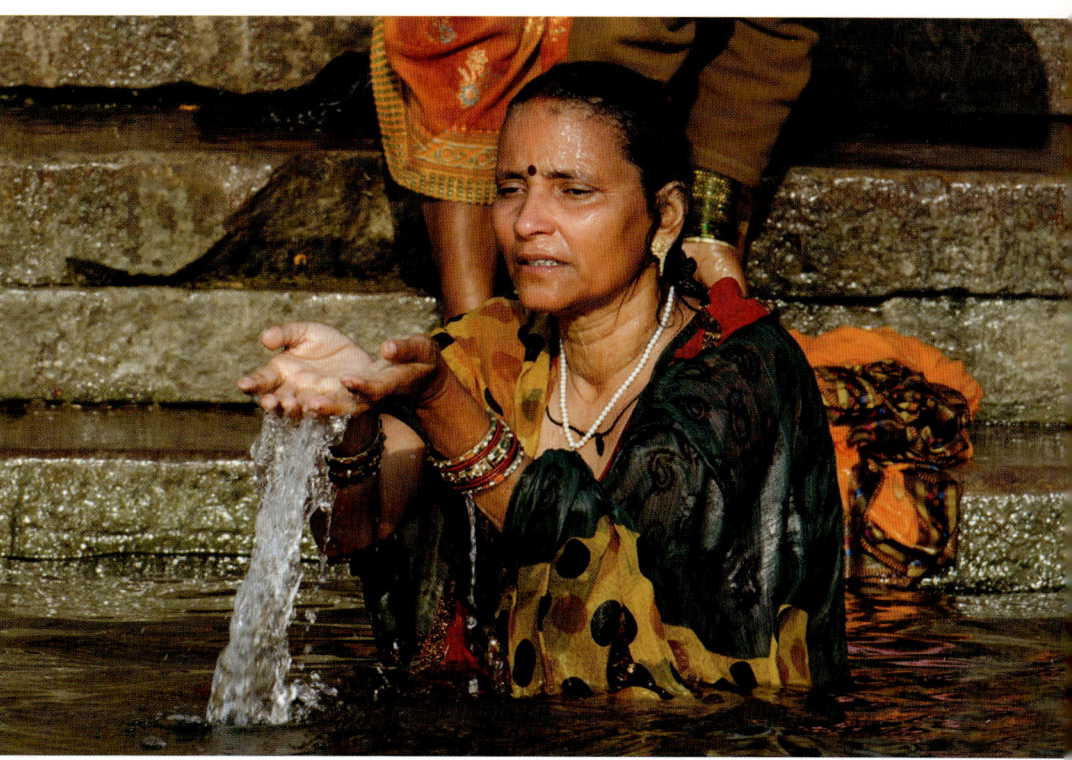

The Largest Religious Festival in the World

Several kilometres before I even reach the Ganges, a barrier is blocking the road. Whistling and wildly gesticulating, a security guard directs me into a car park. I have to leave the car and find another way to get to the river. It is an attempt by the authorities to exercise at least a measure of control over the inrush of the masses. Everyone wants to take part in the Kumbh Mela, the largest religious festival in the world. No one knows exactly how many millions of people will make the pilgrimage to Haridwar in northern India this year – but the Kumbh Mela is always an organizational challenge.

There are a few bicycle rickshaws commuting between the car park and the river. The drivers are quite enterprising, adjusting their prices according to the demand. I patiently wait until it is my turn. In the end I can only travel a short distance, because even the rickshaws cannot get through, and I have to continue on foot. The crowds push through the old alleyways, which get ever narrower as they approach the river. It is stiflingly hot, and the noise increases all the time. Over well-worn steps they descend to the riverbank, to reach the highlight of their pilgrimage – a dip in the holy waters of the Ganges.

A pilgrim needs some patience to take part in a Kumbh Mela. There are various types of this festival, which are held at four sacred sites. The largest is only celebrated every 144 years, with smaller ones every three years. The venue and date depend on the phases of the moon and the position of the stars. But when there is the prospect of bathing in a sacred river at a holy site at the right time, it is all worthwhile.

For Hindus the Ganges is a sacred site, the embodiment of the goddess Ganga. According to Hindu teaching, the water has the power of spiritual cleansing. By bathing in the Ganges the faithful can wash away their sins. The water is also used during everyday religious rituals. The belief in its positive properties is unbroken even today – which is why it is also used for drinking.

This is surprising, considering that the Ganges is one of the most polluted rivers on earth. At its source, near the snowfields of the Himalayas, the water is fresh and pure. But on the densely populated Indo-Gangetic Plain, where huge amounts of untreated sewage are discharged into the river, the pollution rapidly increases. With high water temperatures, the bacteria reach dangerous concentrations. The river around the town of Varanasi is especially contaminated.

Such earthly problems do not diminish the outstanding religious importance of the place. Varanasi is one of the oldest continuously inhabited cities in India and also one of the most sacred sites in Hinduism. The Hindu teachings on karma and reincarnation have many followers even today. It is believed that if people conduct themselves in a positive way, they can improve their karma and increase their chances of a better next life. Through enlightenment, a believer can break through the eternal cycle of birth, death and rebirth. But even after years of meditation and asceticism the outcome remains uncertain. Varanasi offers hope, because the city promises salvation from earthly torments. If you die here, and are cremated and have your ashes scattered on the river, you will leave the cycle of reincarnations behind you.

Every morning thousands of Hindus pray on the banks of the river. As a sign of their devotion they offer blossoms and floating candles to the holy waters. Many elderly people are among the devotees. Should they actually die here, it would be their salvation.

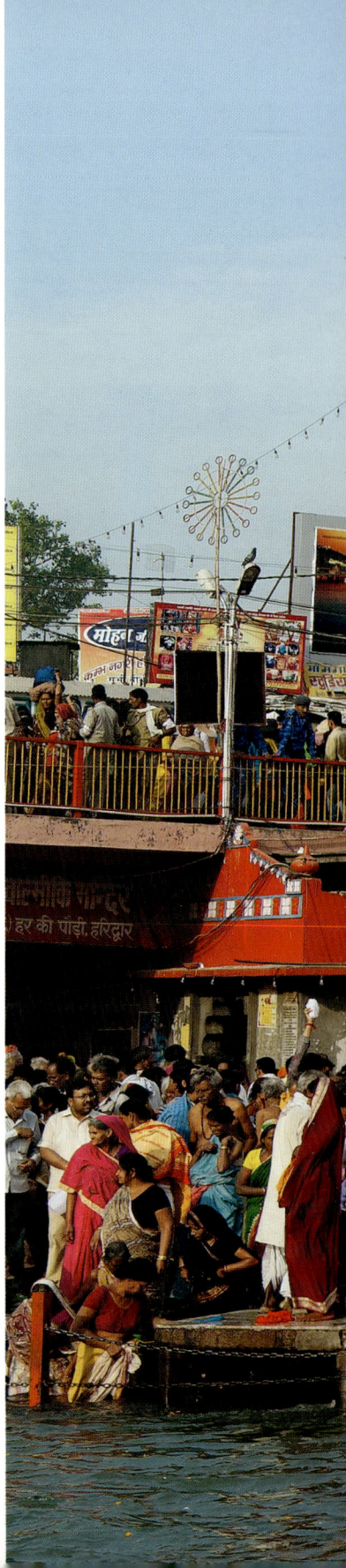

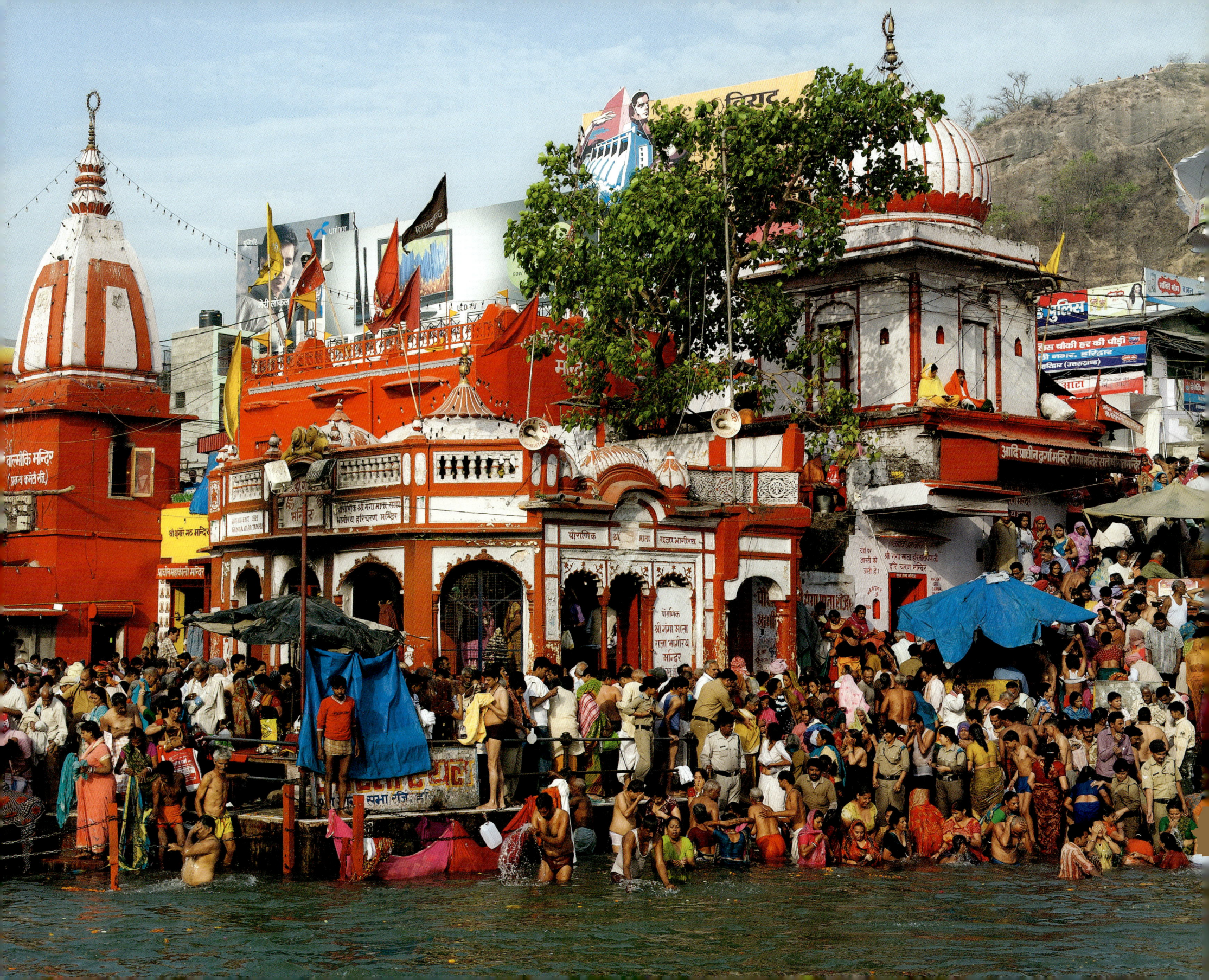

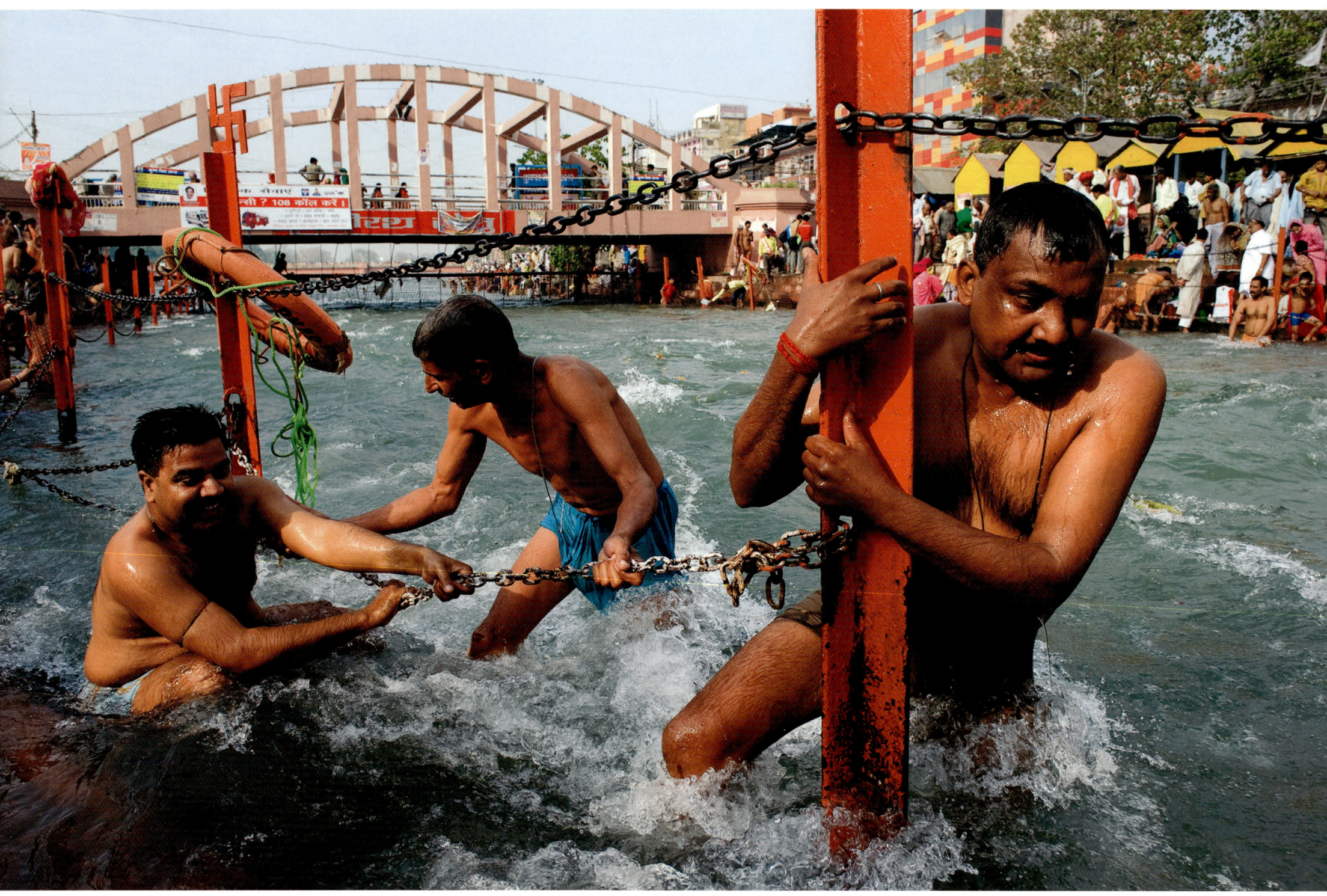

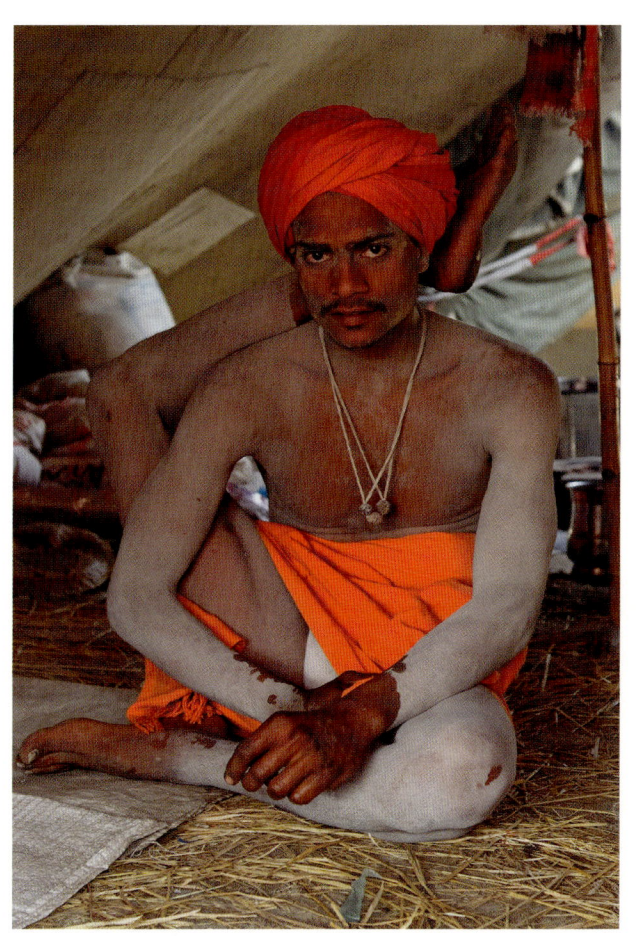
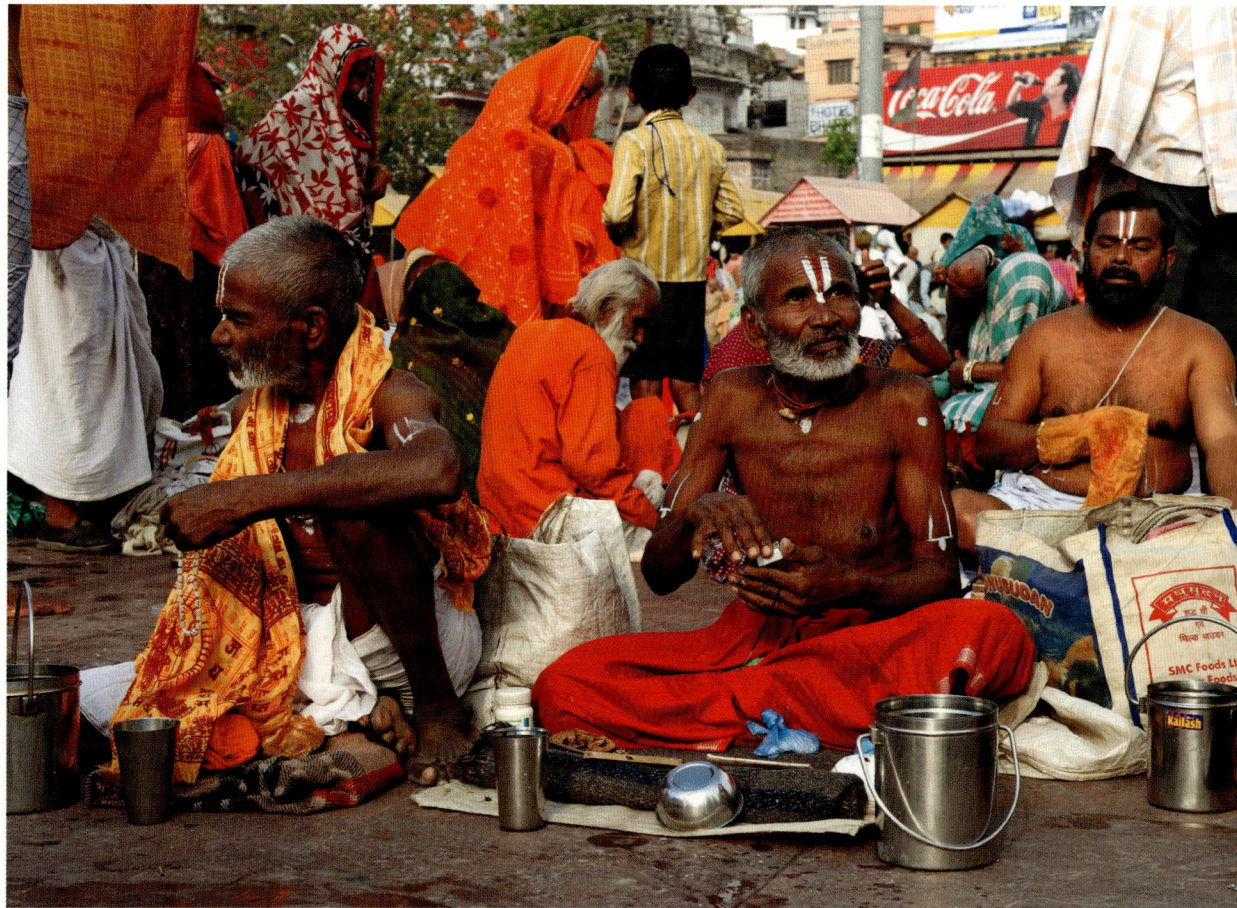

Every few years, the religious festival of Kumbh Mela takes place in Haridwar. As the city is situated comparatively close to the source of the Ganges, the river here is cold and the current is strong. So the faithful find something to hold onto while they bathe in the holy waters. Pilgrims have set up camp throughout the area, with some fakirs among them.

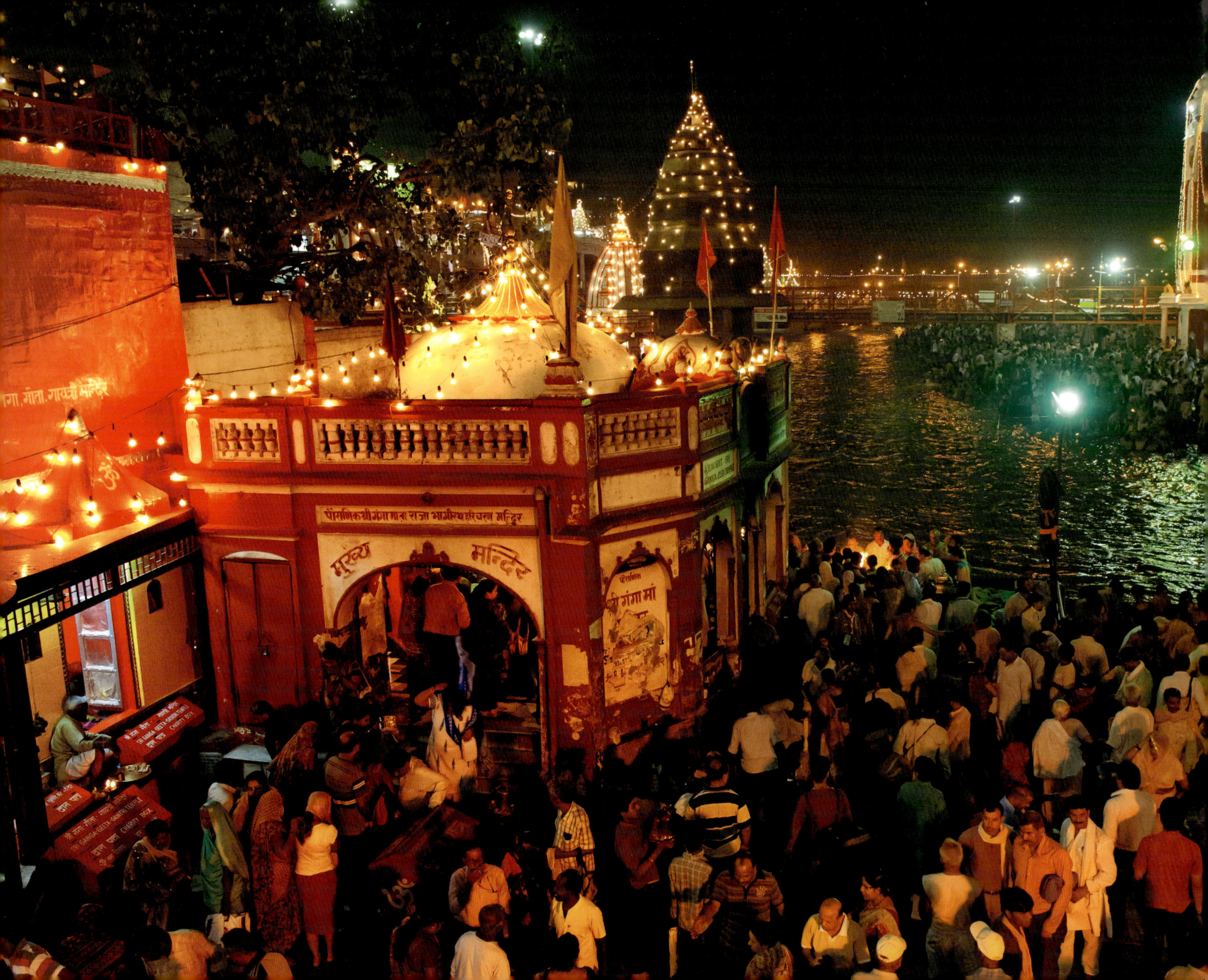

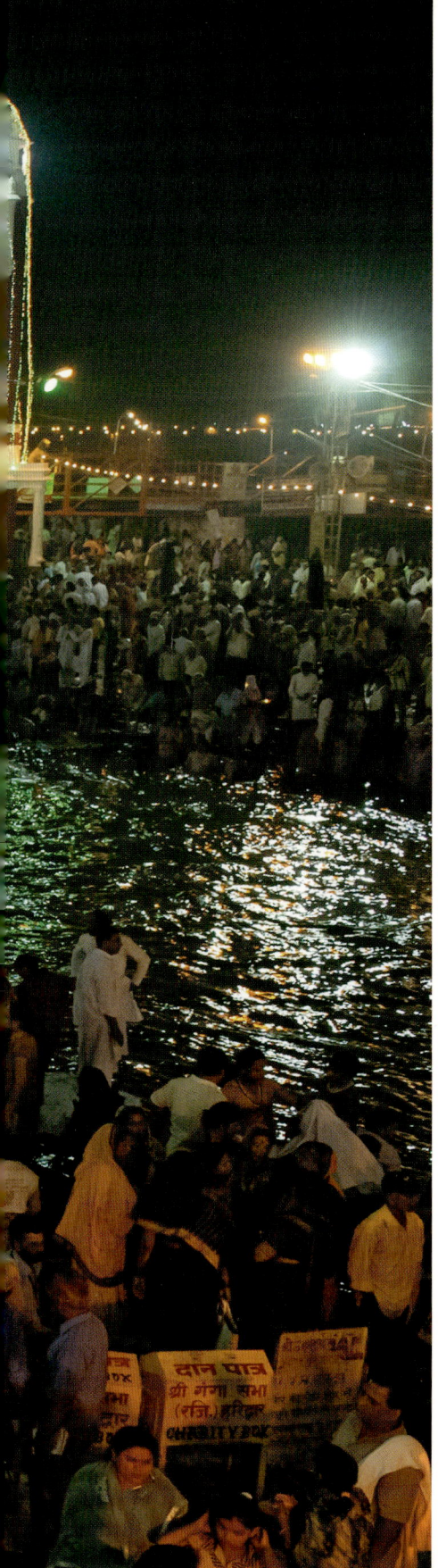

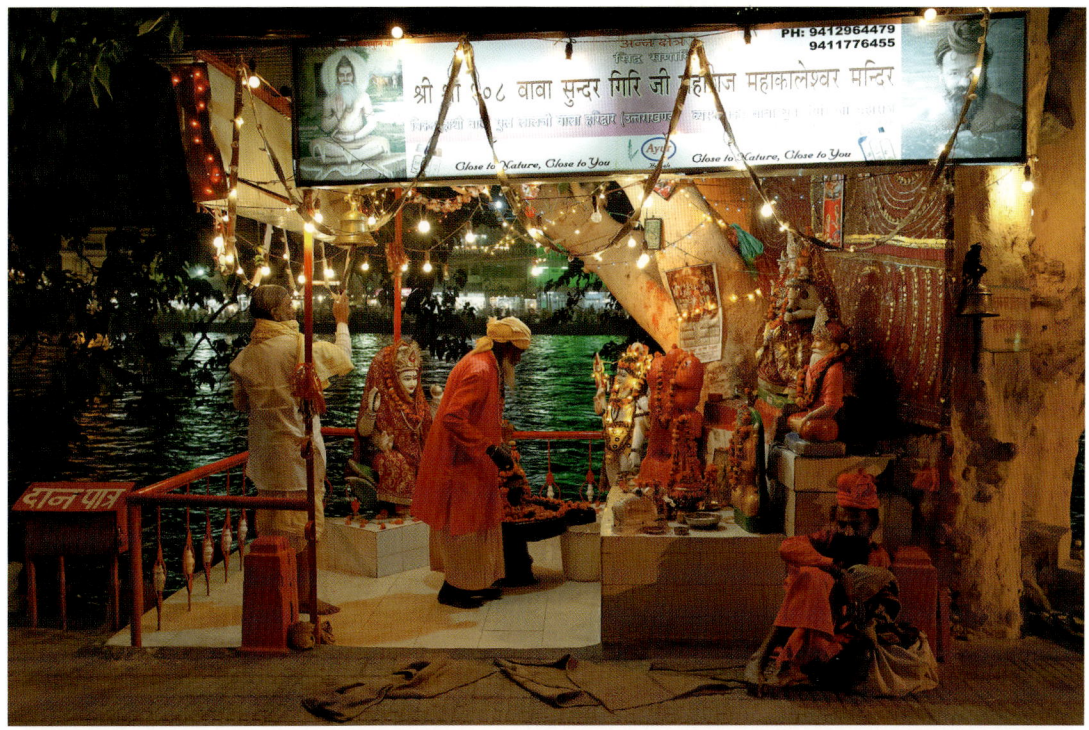

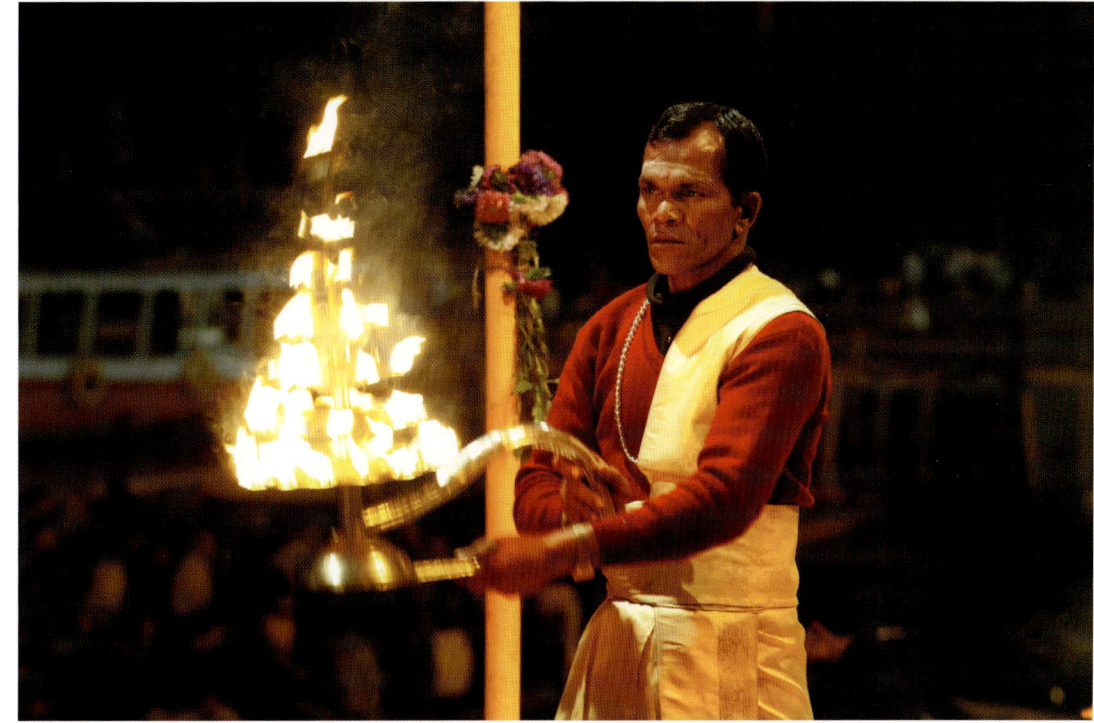

During the Kumbh Mela several million pilgrims crowd onto the banks of the Ganges. Many of the religious ceremonies take place in the evening, when it is cooler.

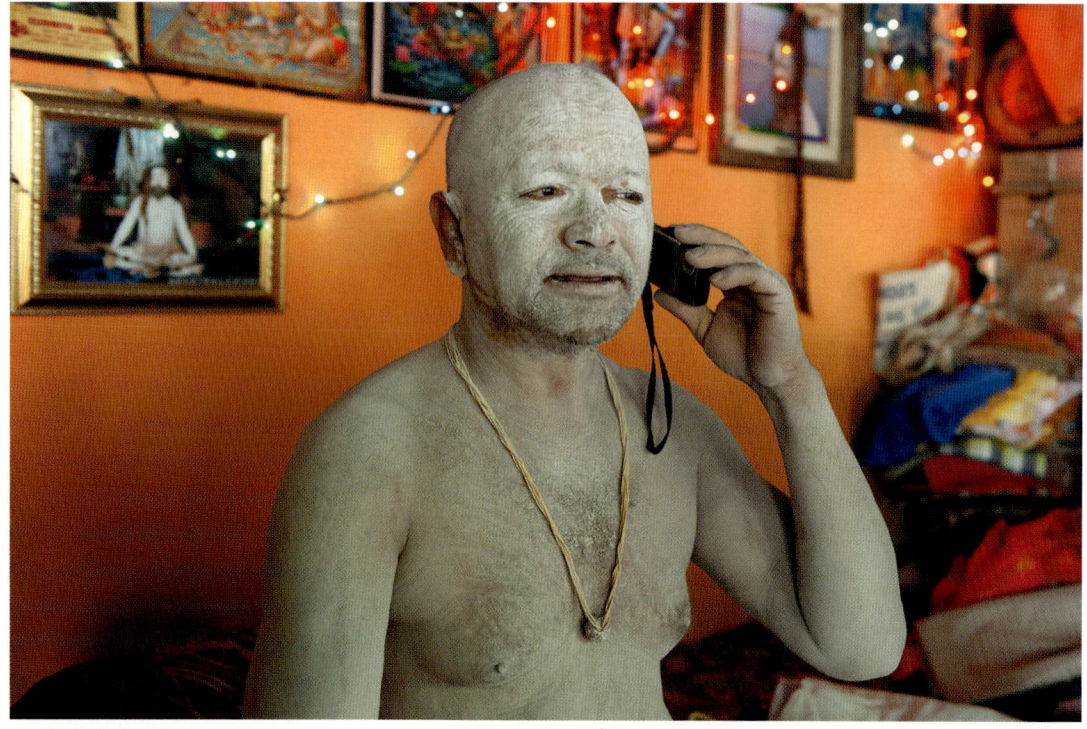

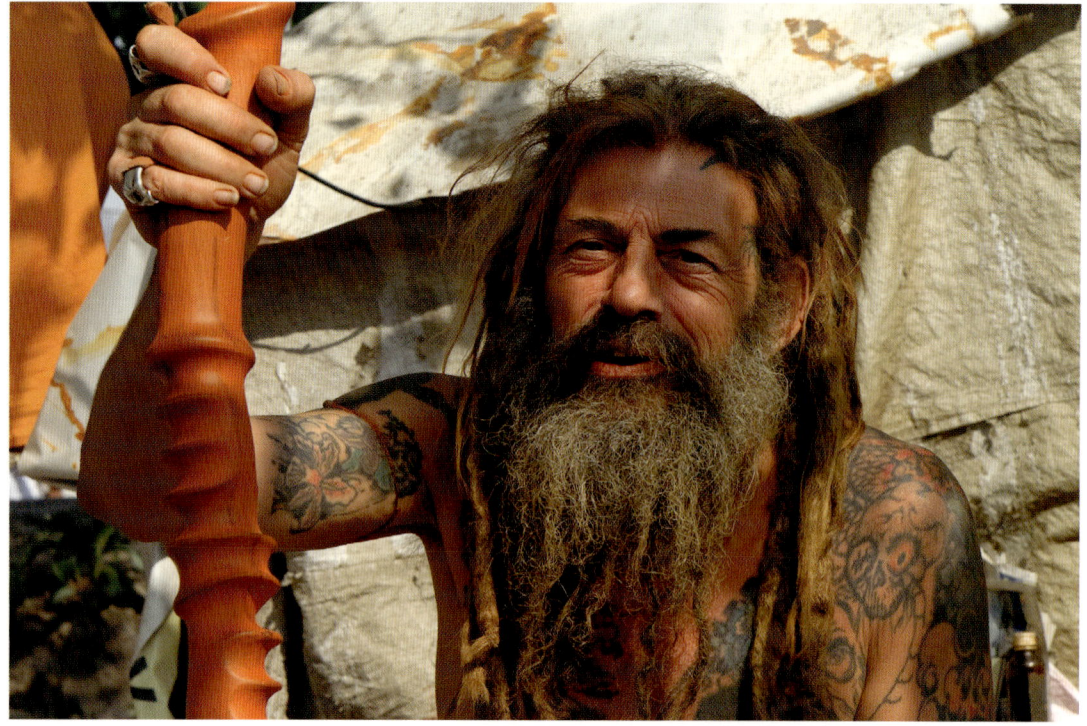

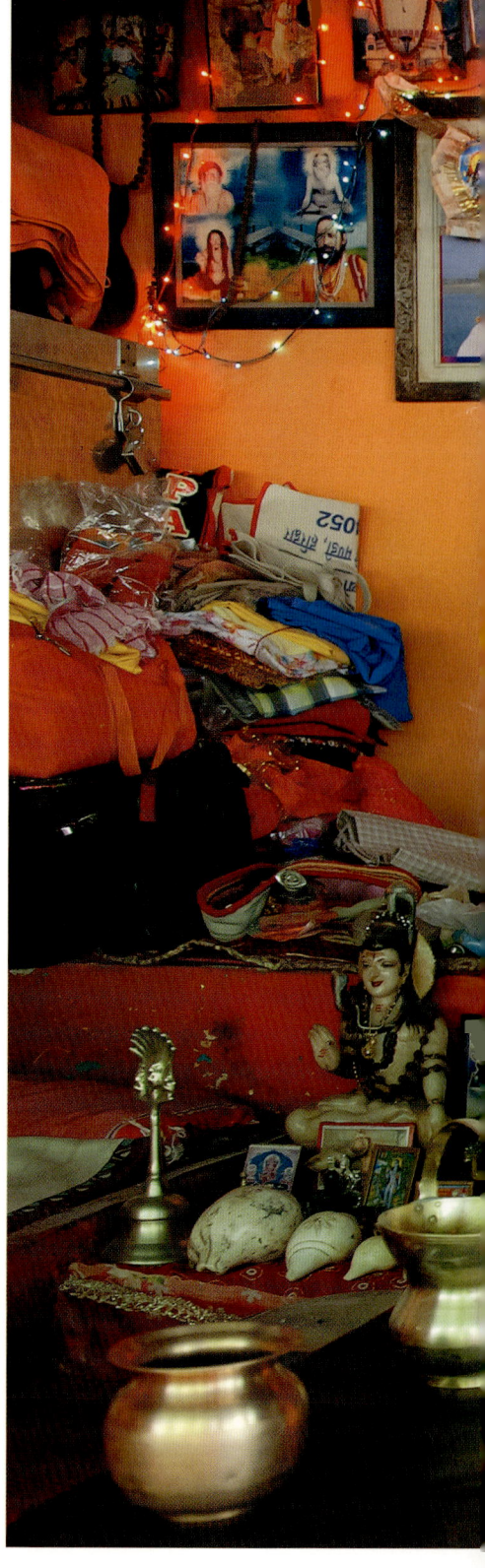

Some of the sadhus, the holy men of India, lead quite a pleasant life. The festival is high season for them. Devotees can get a blessing for a few rupees (right). But modern times are changing the ways of the sadhus, too, and these days the holy men keep in touch by mobile phone. There are even some incomers among them. John came to India from Britain during the hippie era and stayed. Today, he travels the country as a sadhu (left).

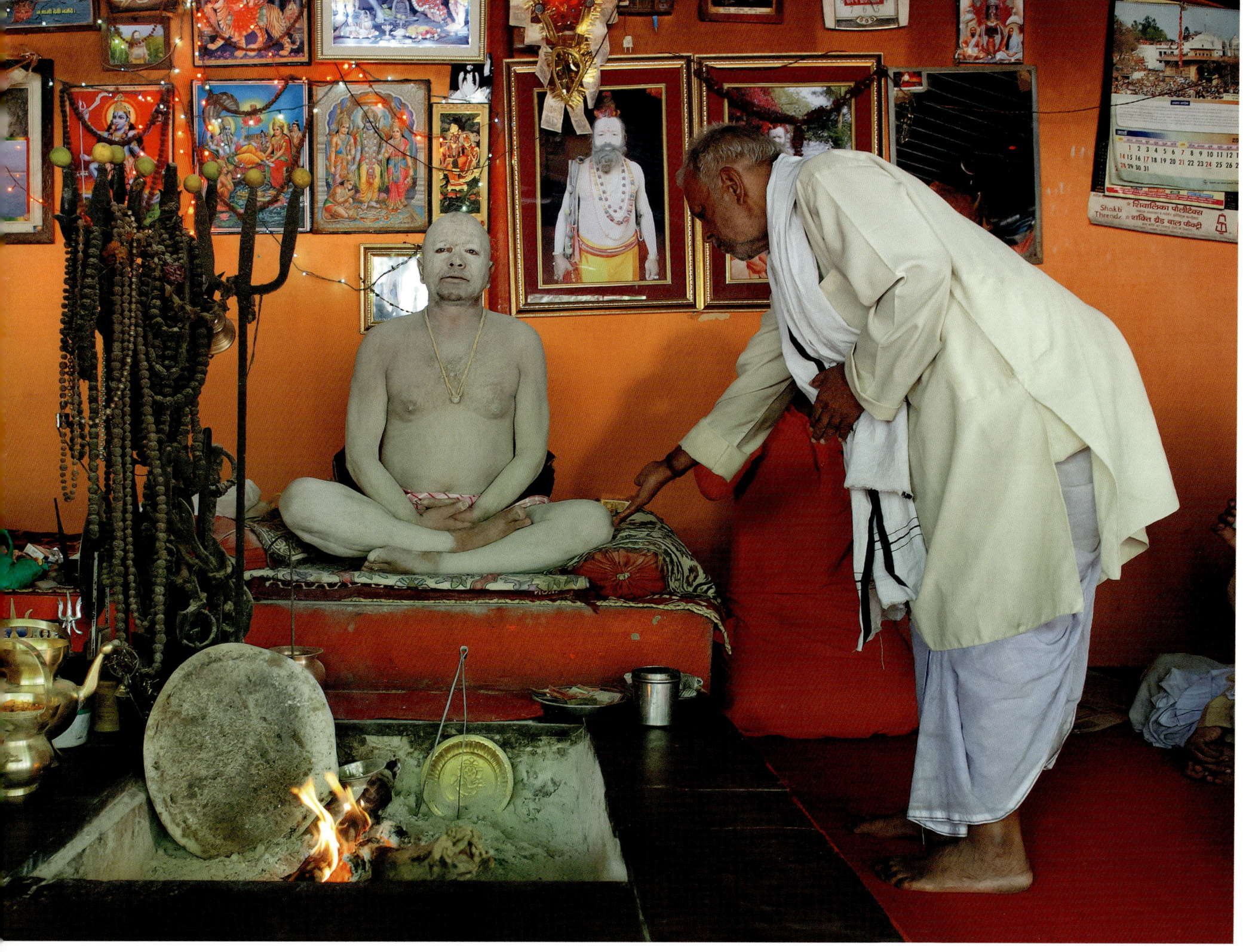

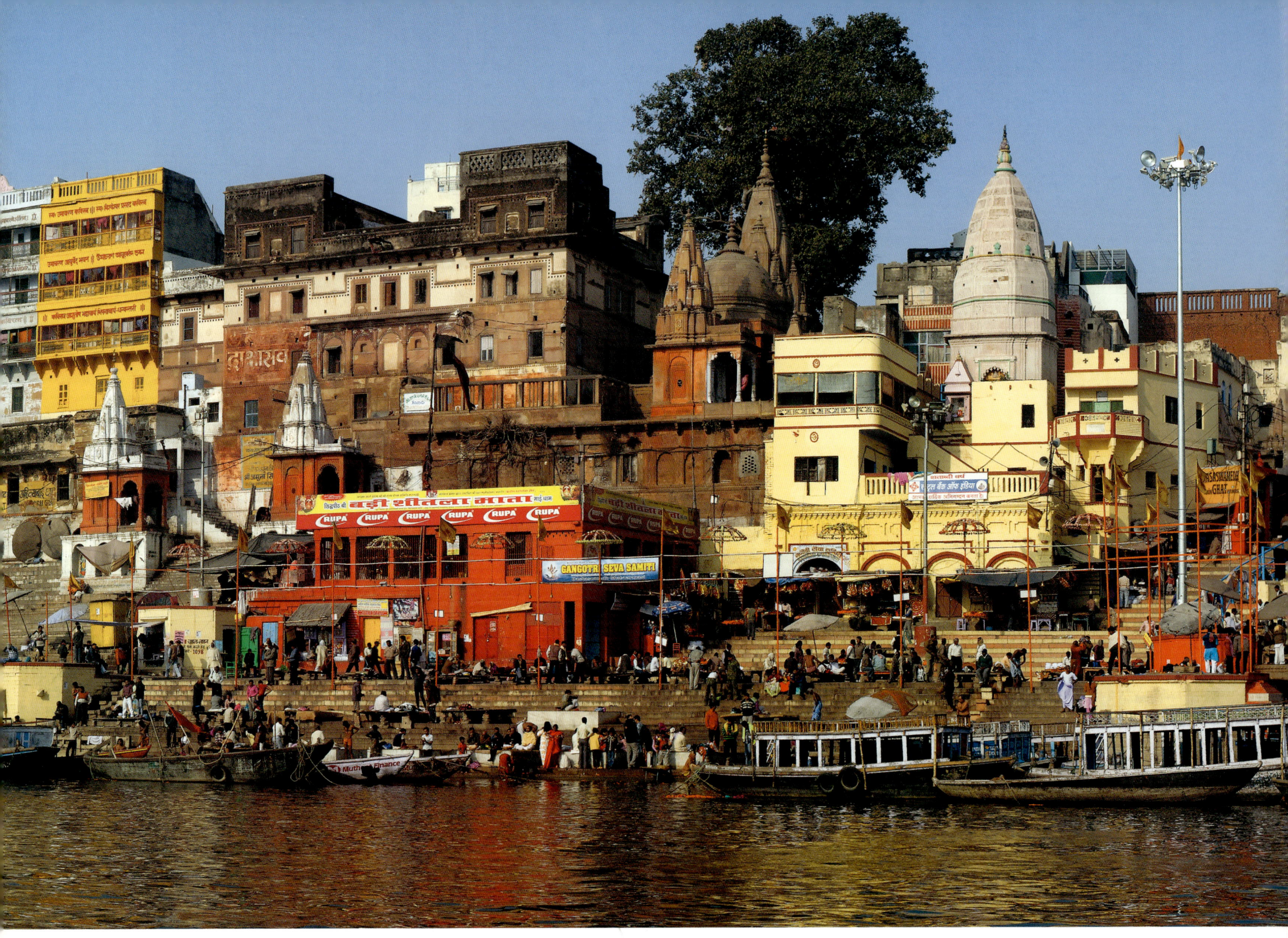

Varanasi, where thousands of believers come to bathe in the Ganges every day, is considered by many to be one of the most 'Indian' cities of the country.

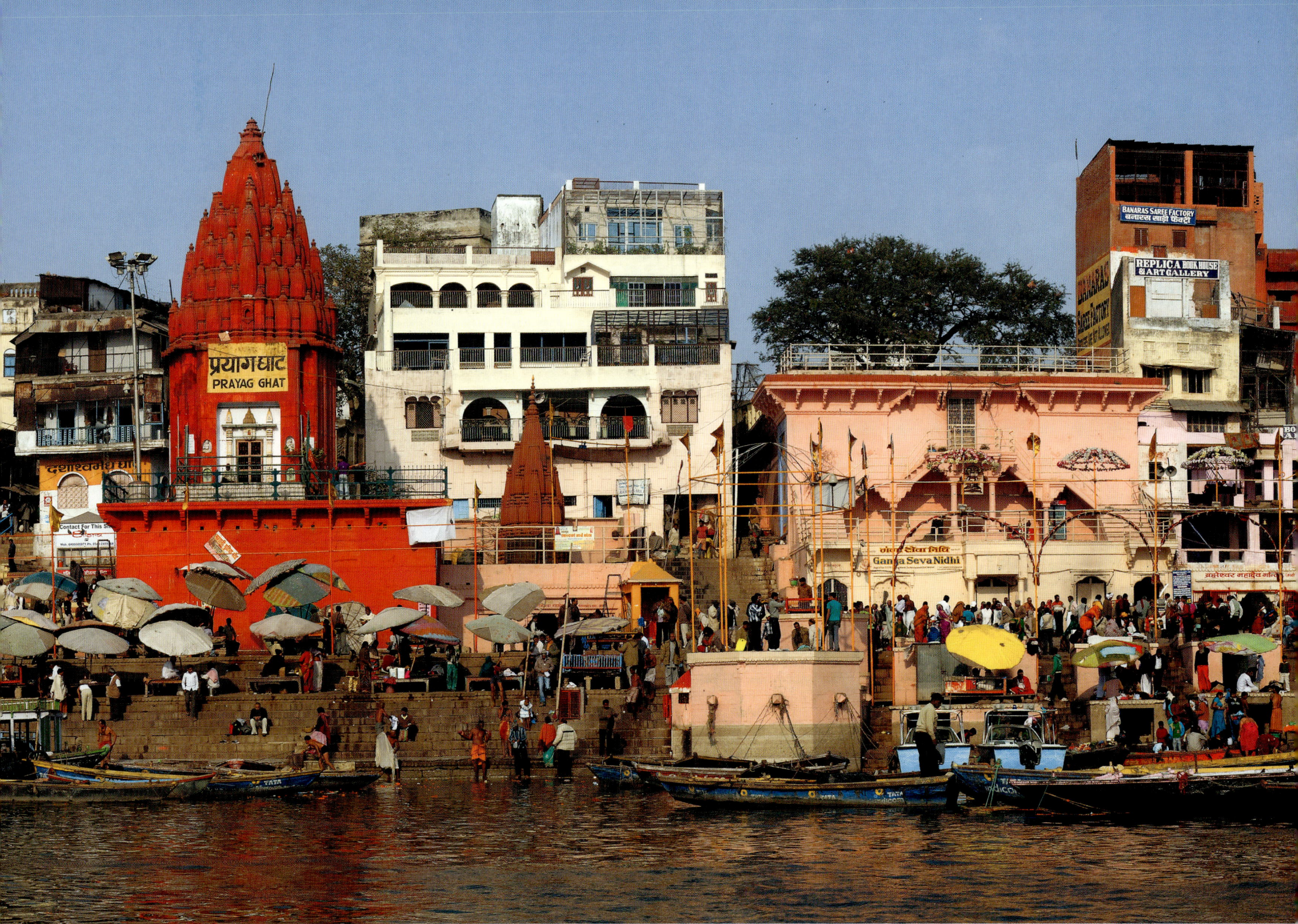

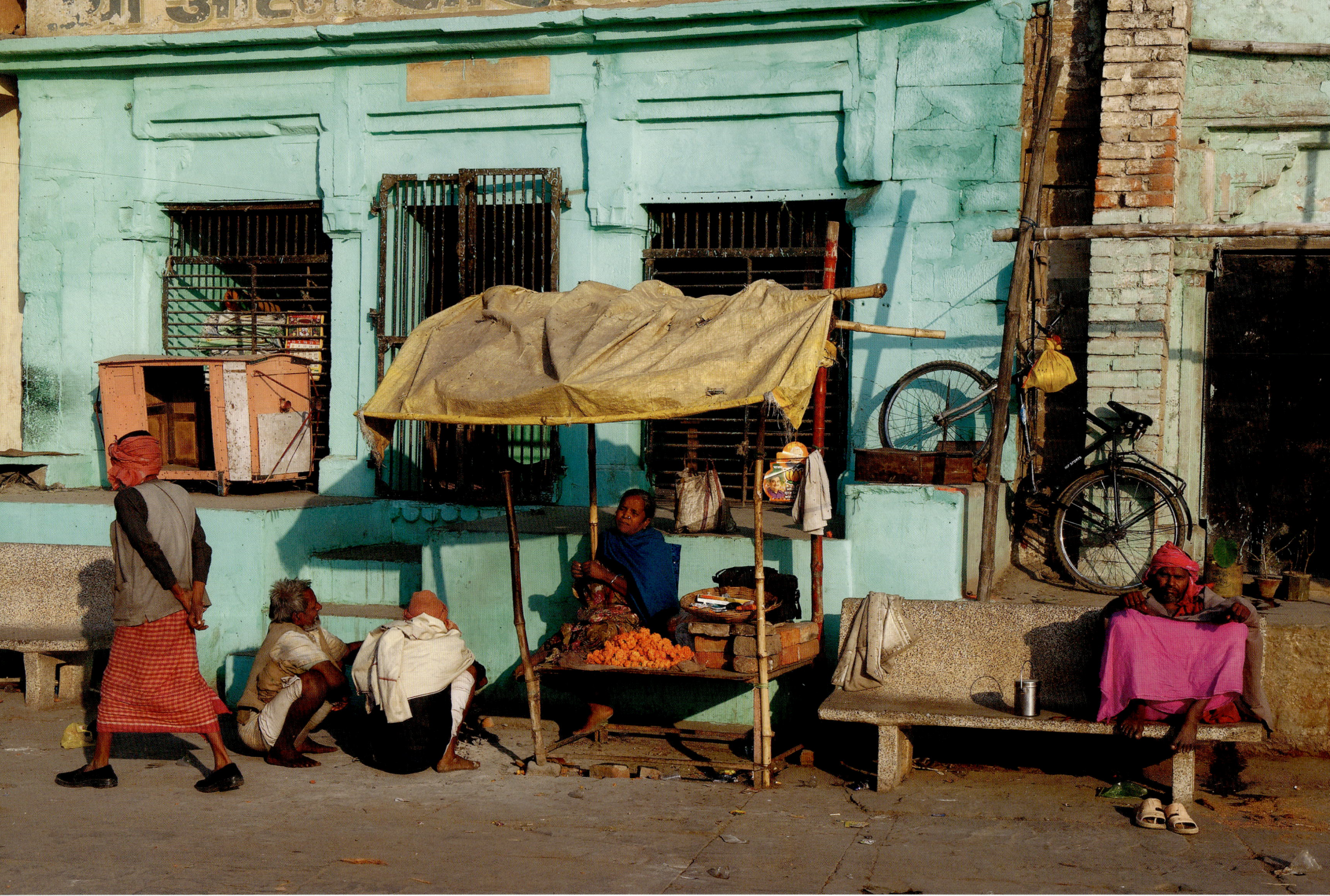

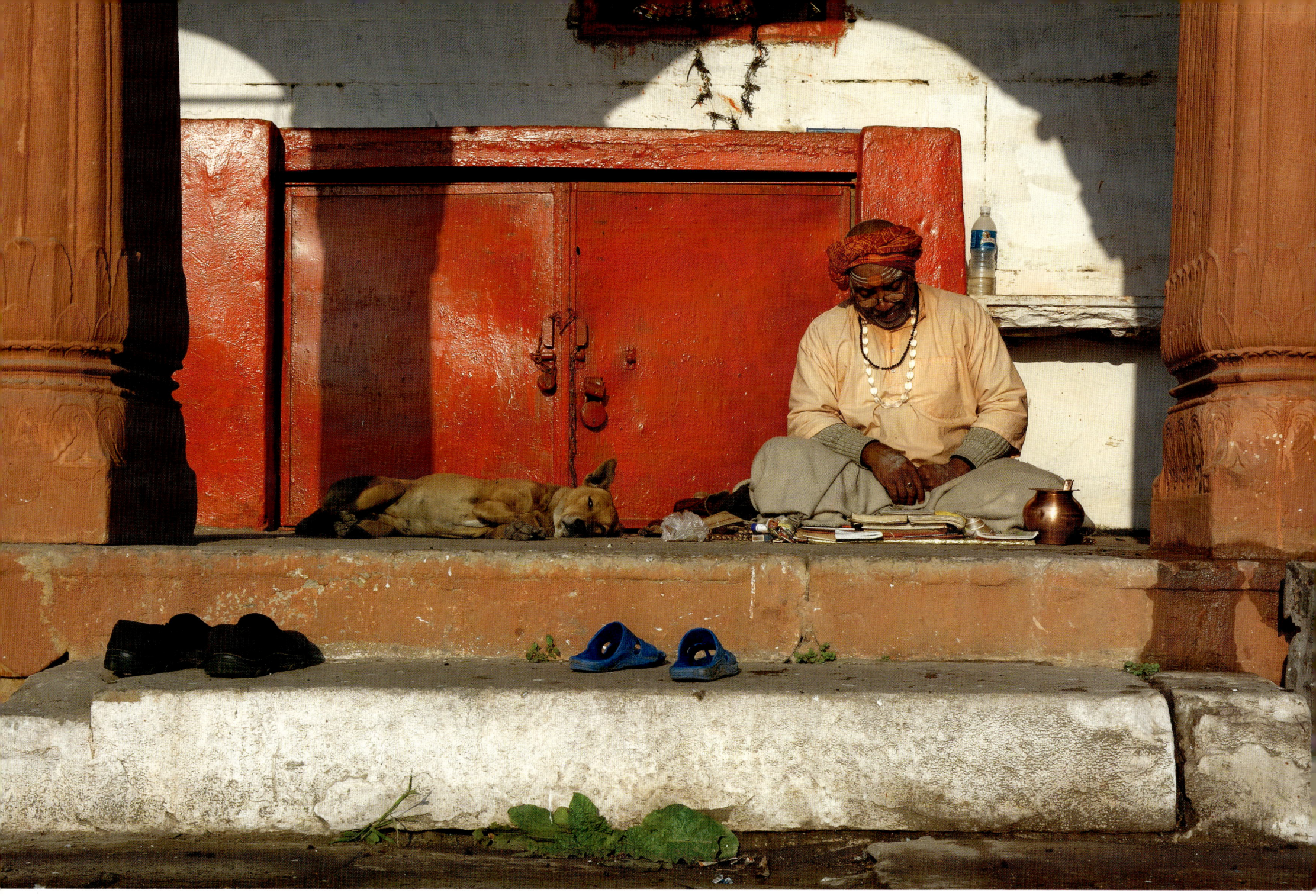

The aura of the sacred site inspires a devout Hindu to study the religious texts. He lives in one of the oldest continuously inhabited cities of India – though an average street scene in Varanasi does little to reveal the age of the city, or its rich history.

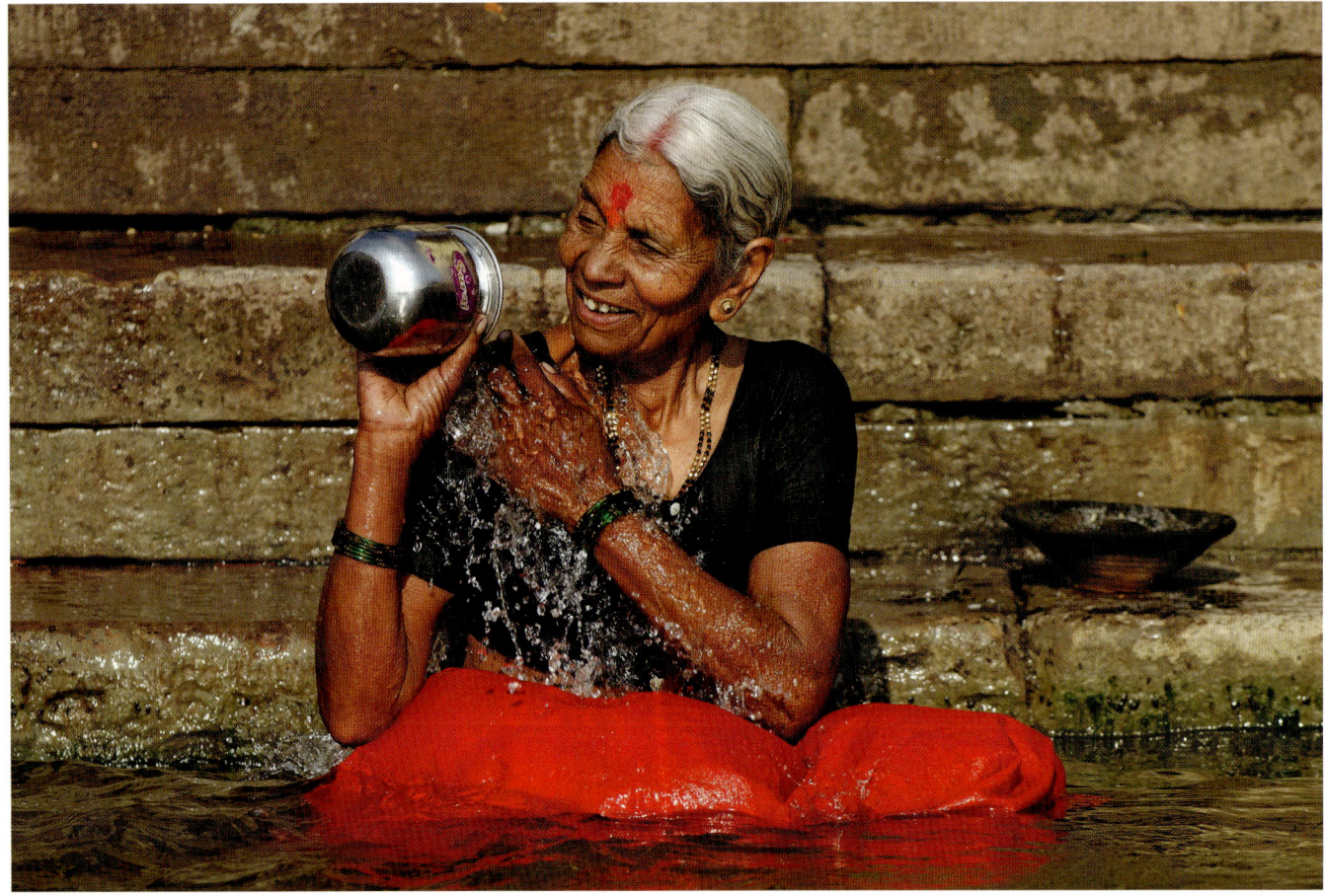

With water from the Ganges, it is believed, you can
cleanse not only your laundry but also your soul.

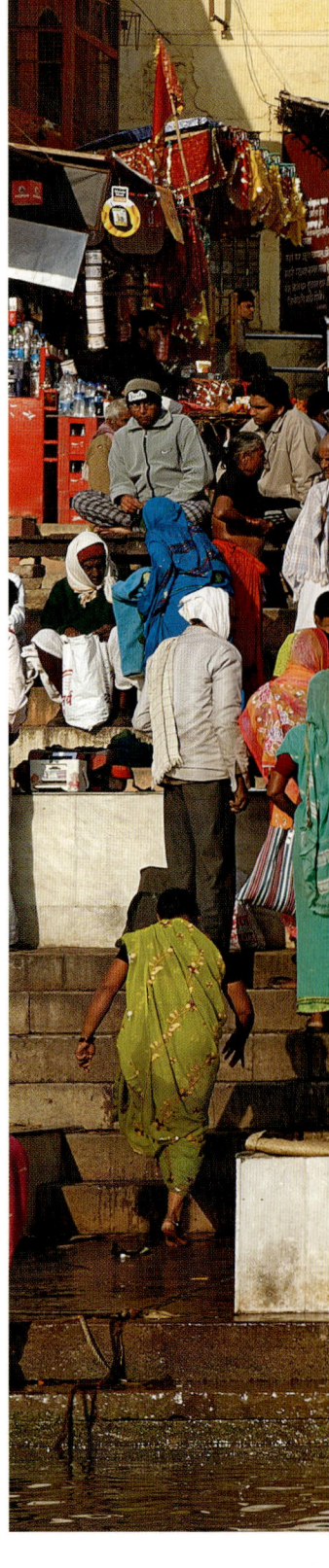

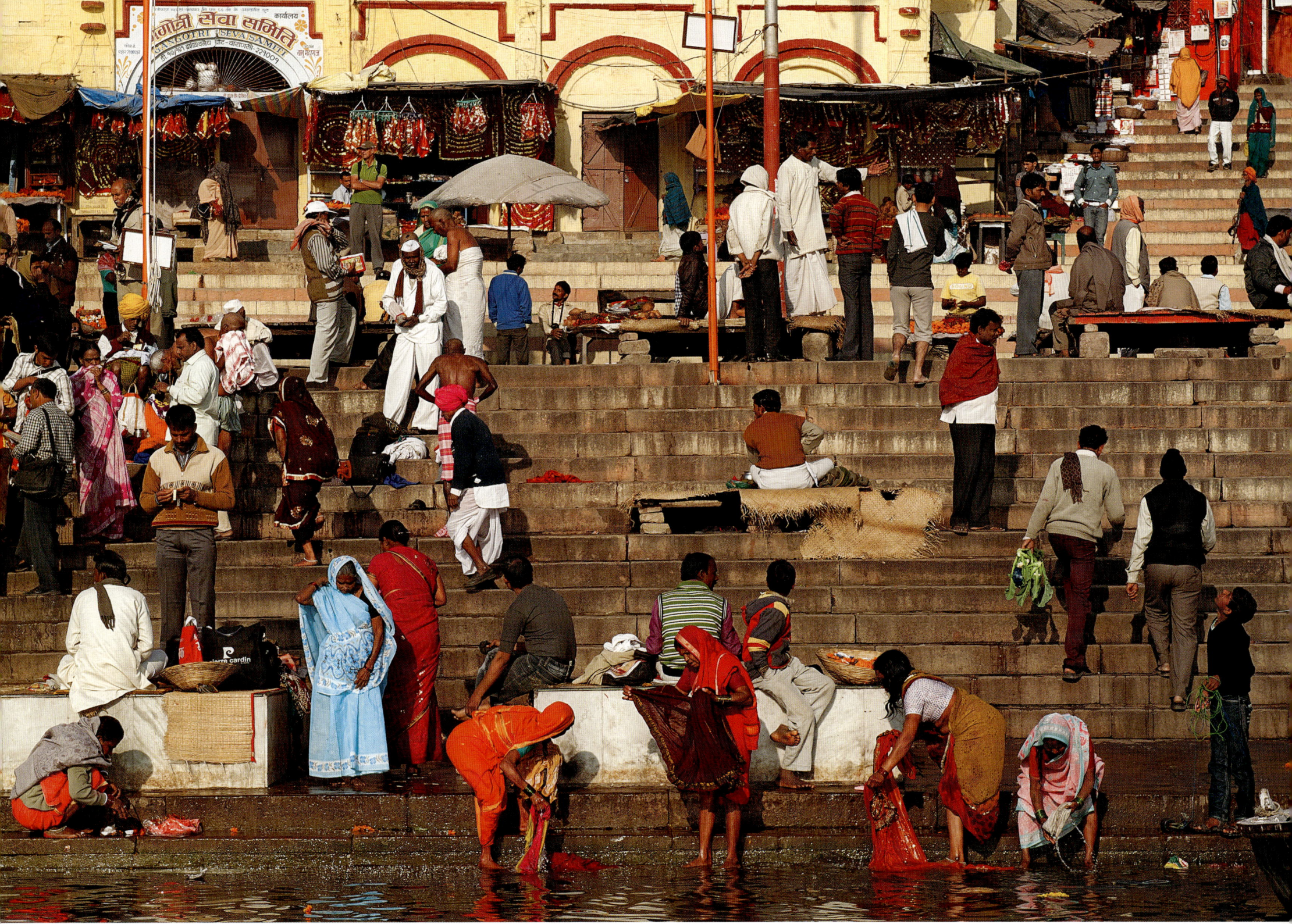

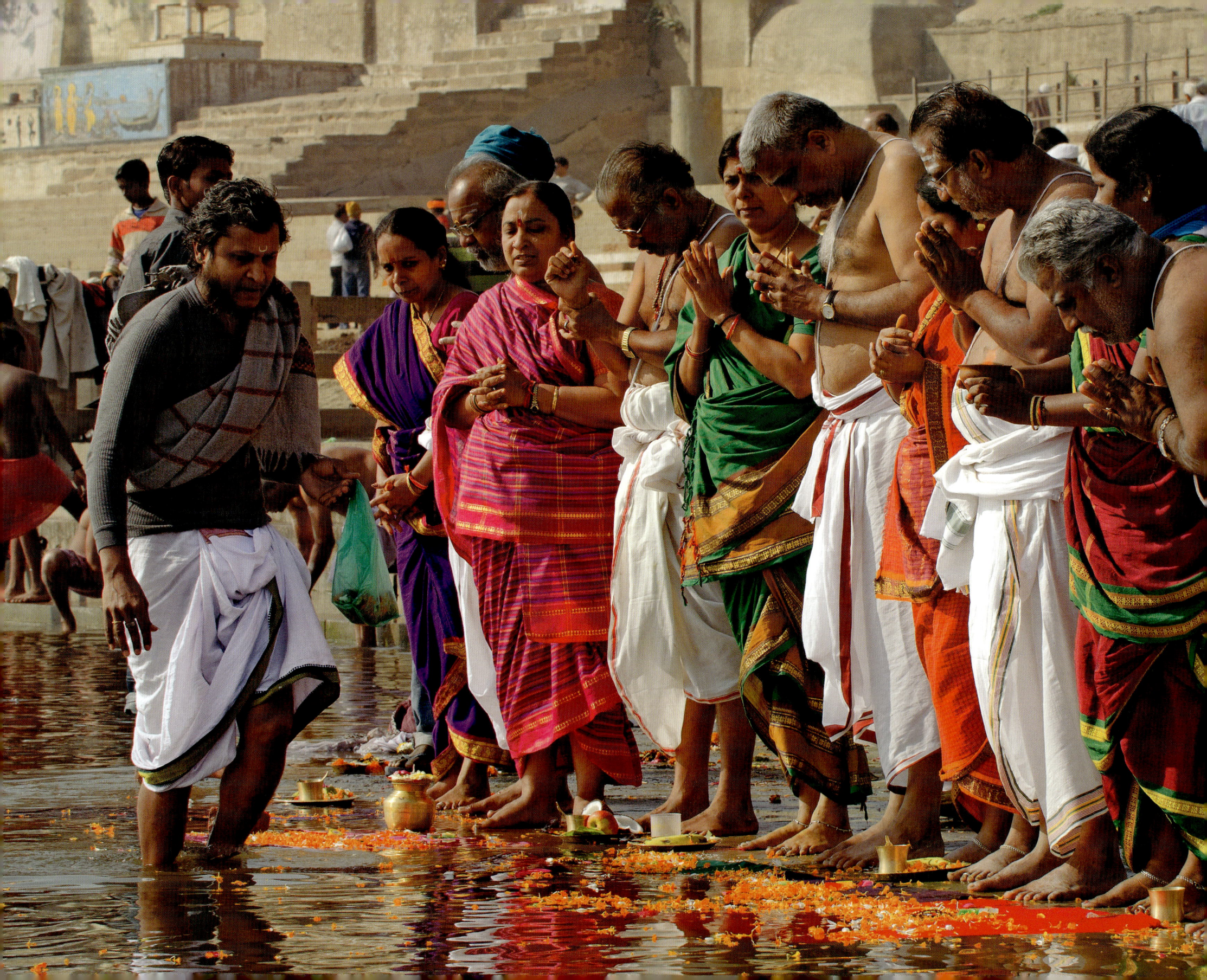

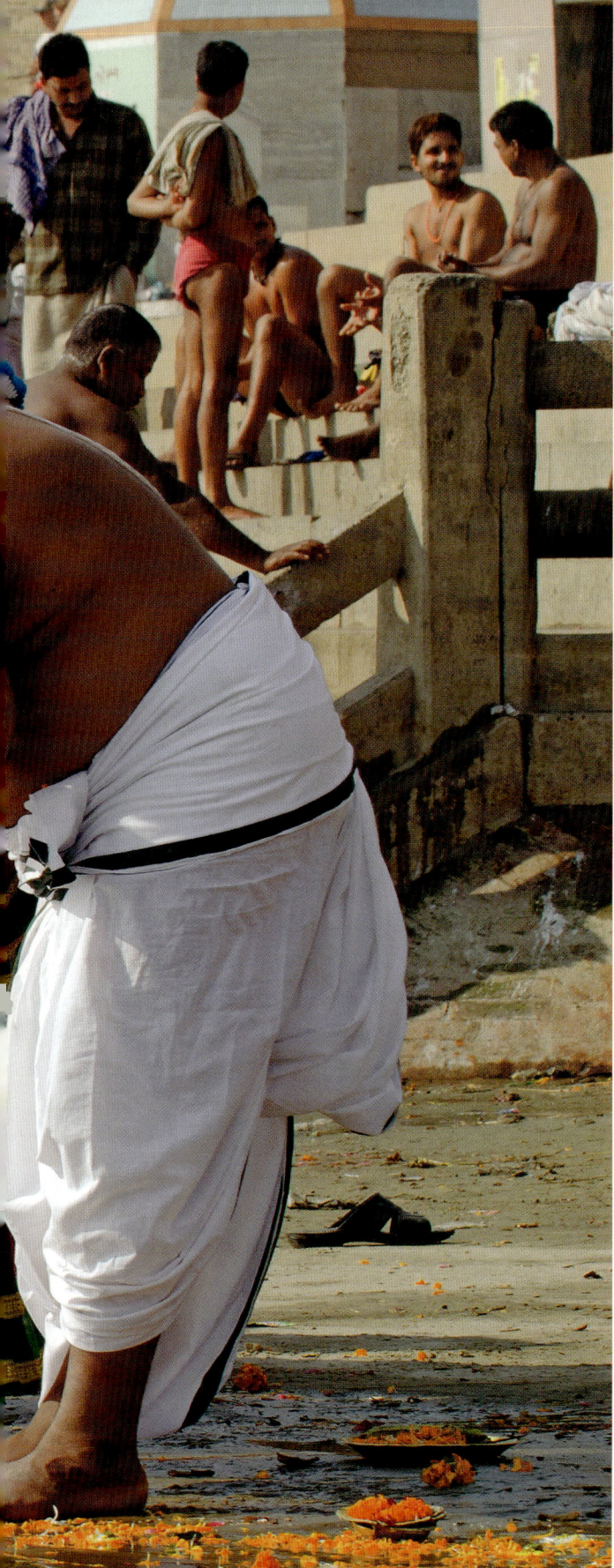

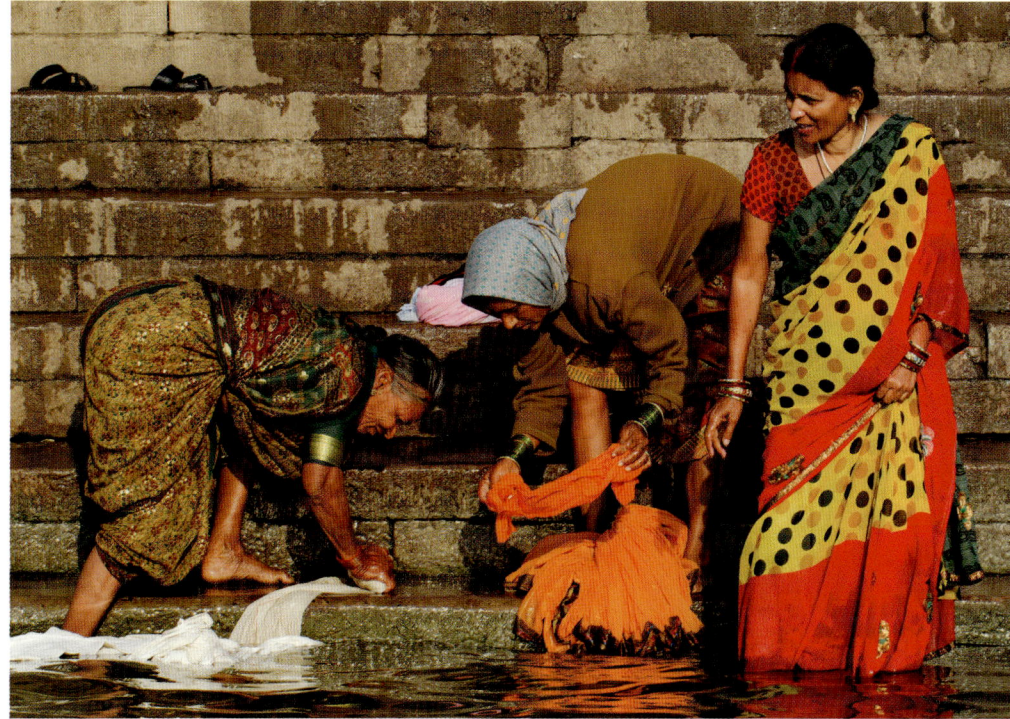

During a Hindu ceremony, blossoms and fruits are presented to the Ganges. Such offerings are for sale anywhere, even on the river itself. The spiritual cleansing takes place right next to where the laundry is washed.

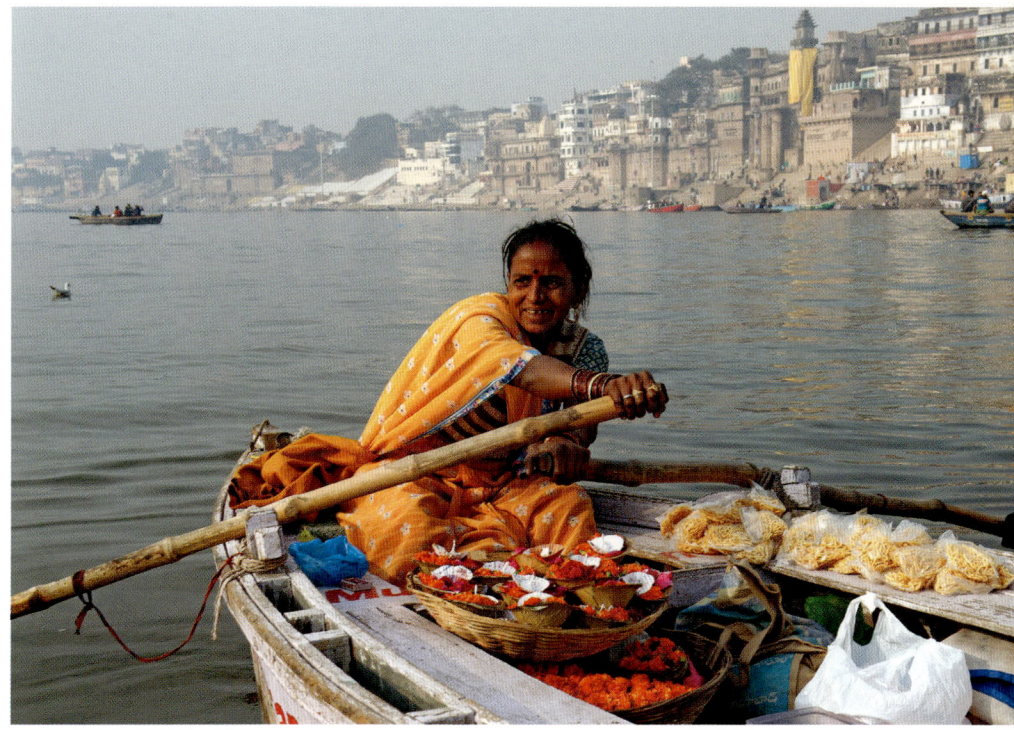

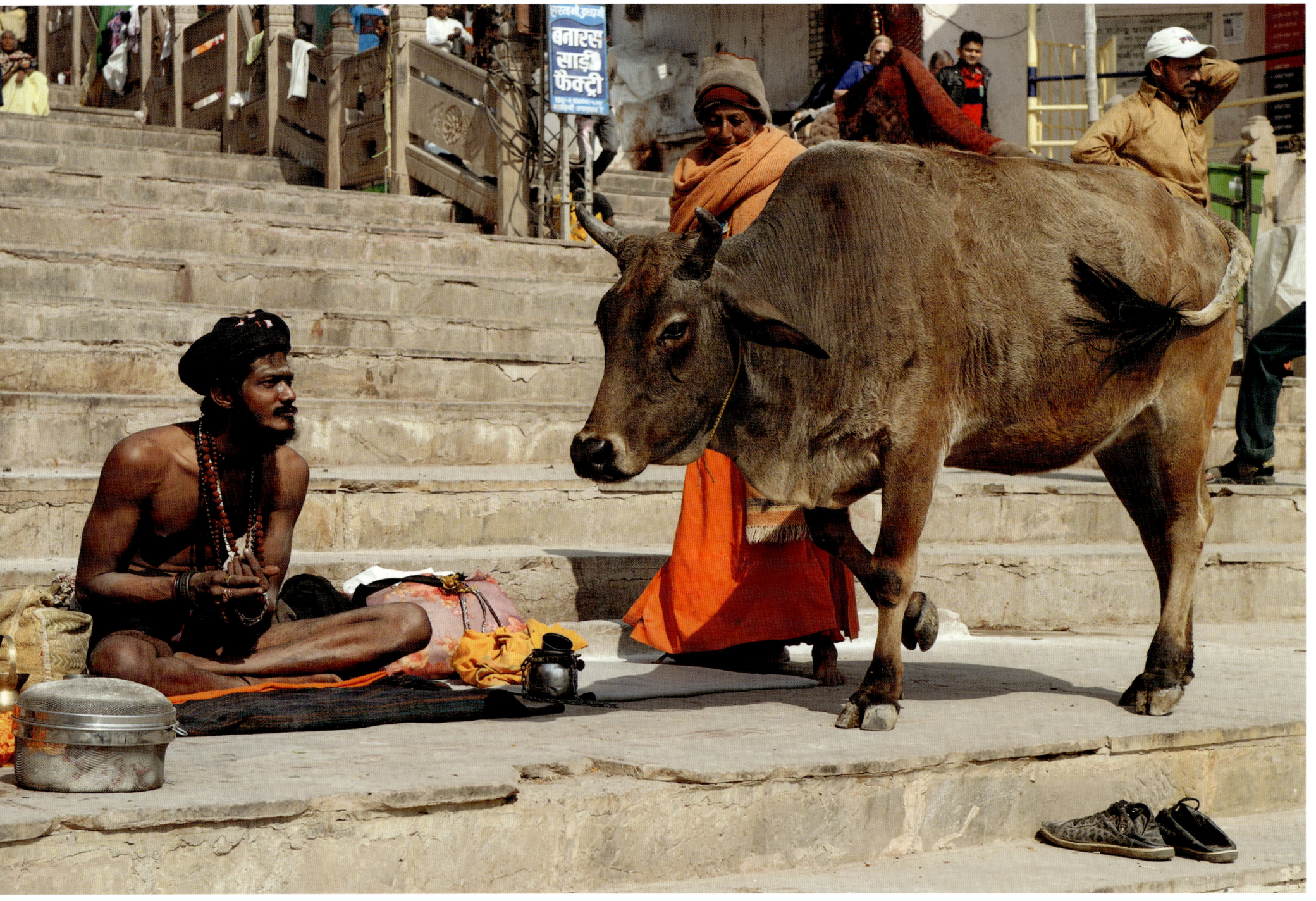

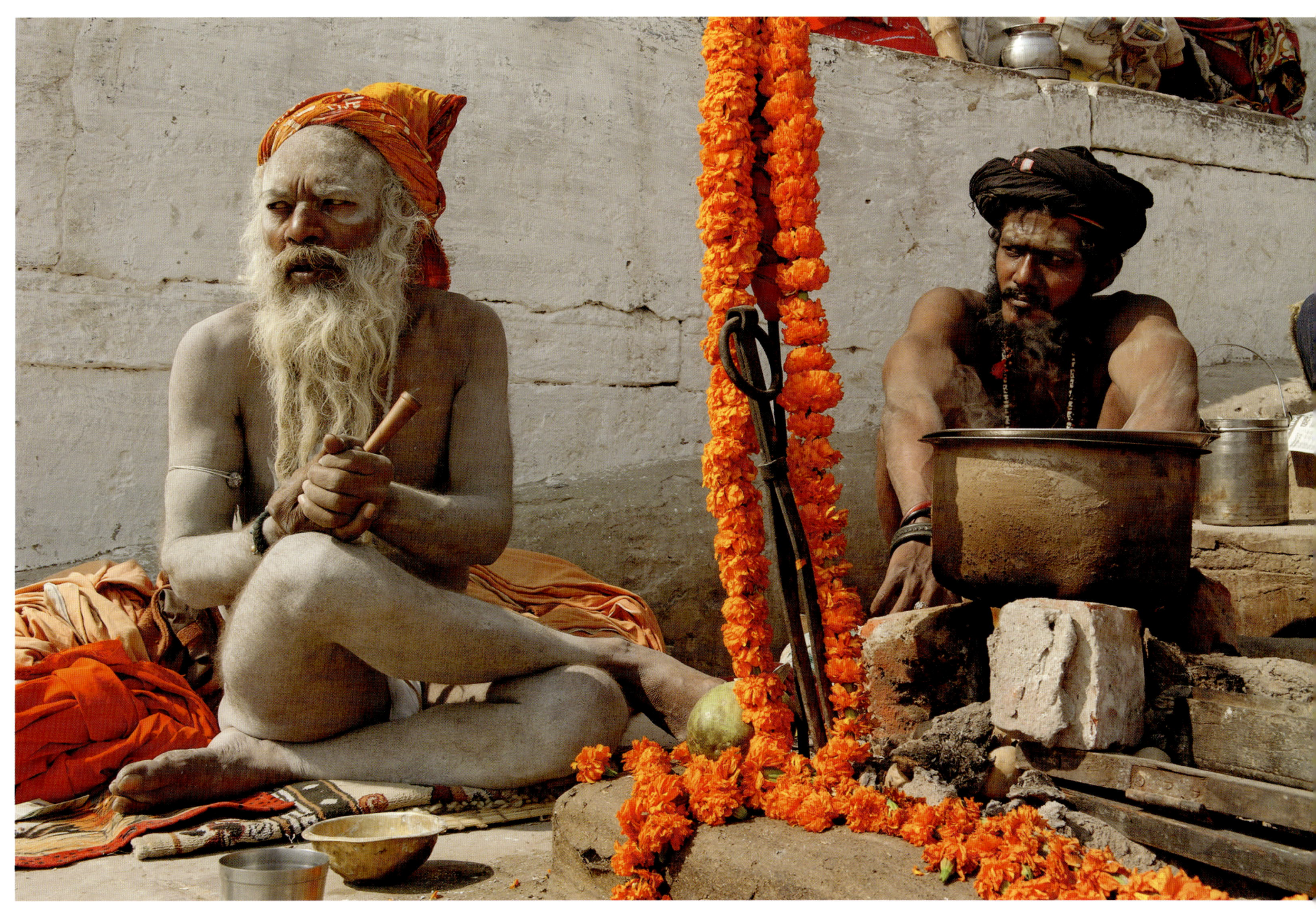

The old holy men, the sadhus, often have novices in their entourage. Respectful dealings with other creatures, such as sacred cows, are a matter of course.

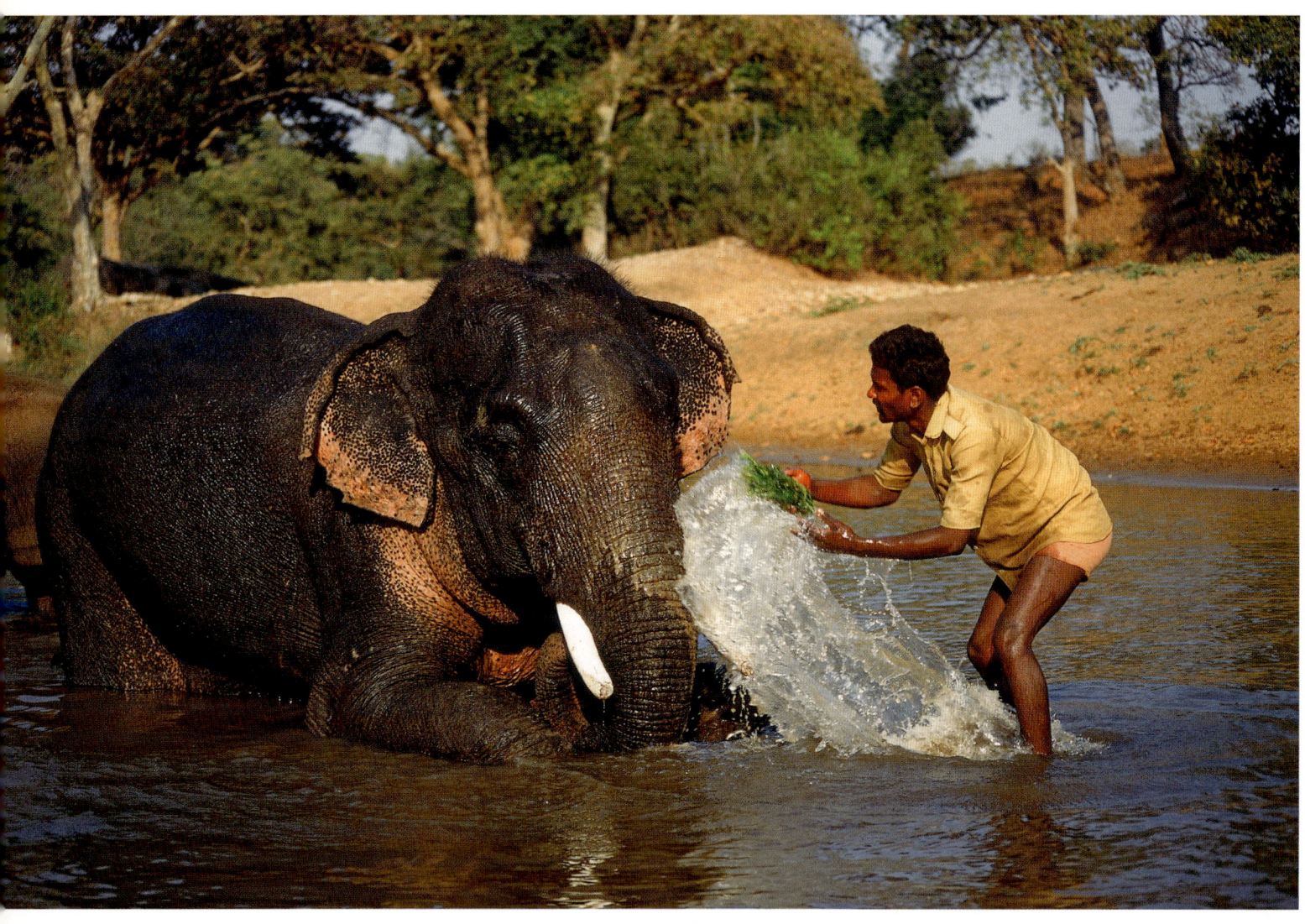

Religious Hindus believe that all forms of life are sacred. They reject violence, and many believers are vegetarians. They respect their fellow creatures, domestic and wild animals alike. A working elephant is helped with its skin care, and joining a group of water buffalos in their bath is a lot of fun.

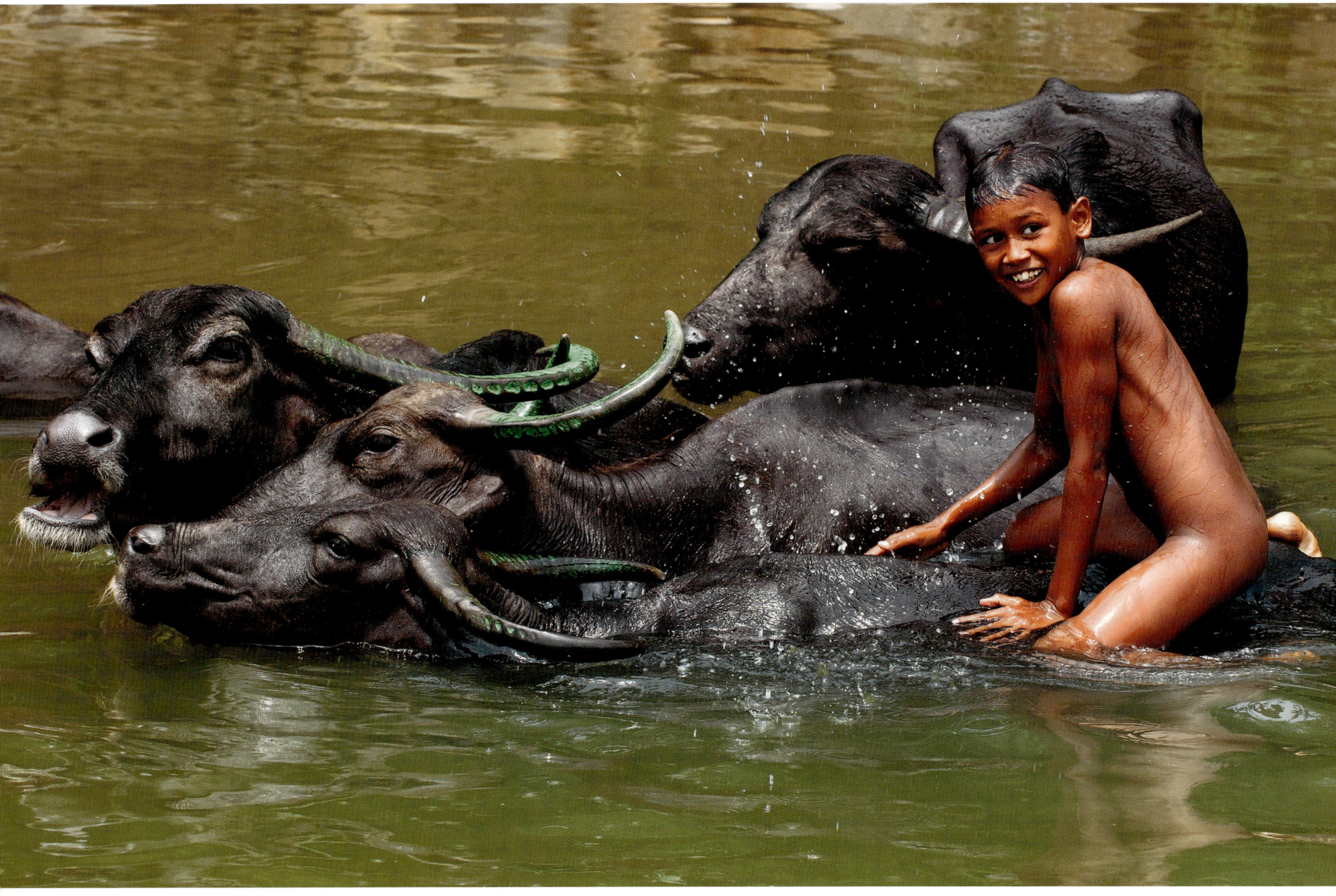

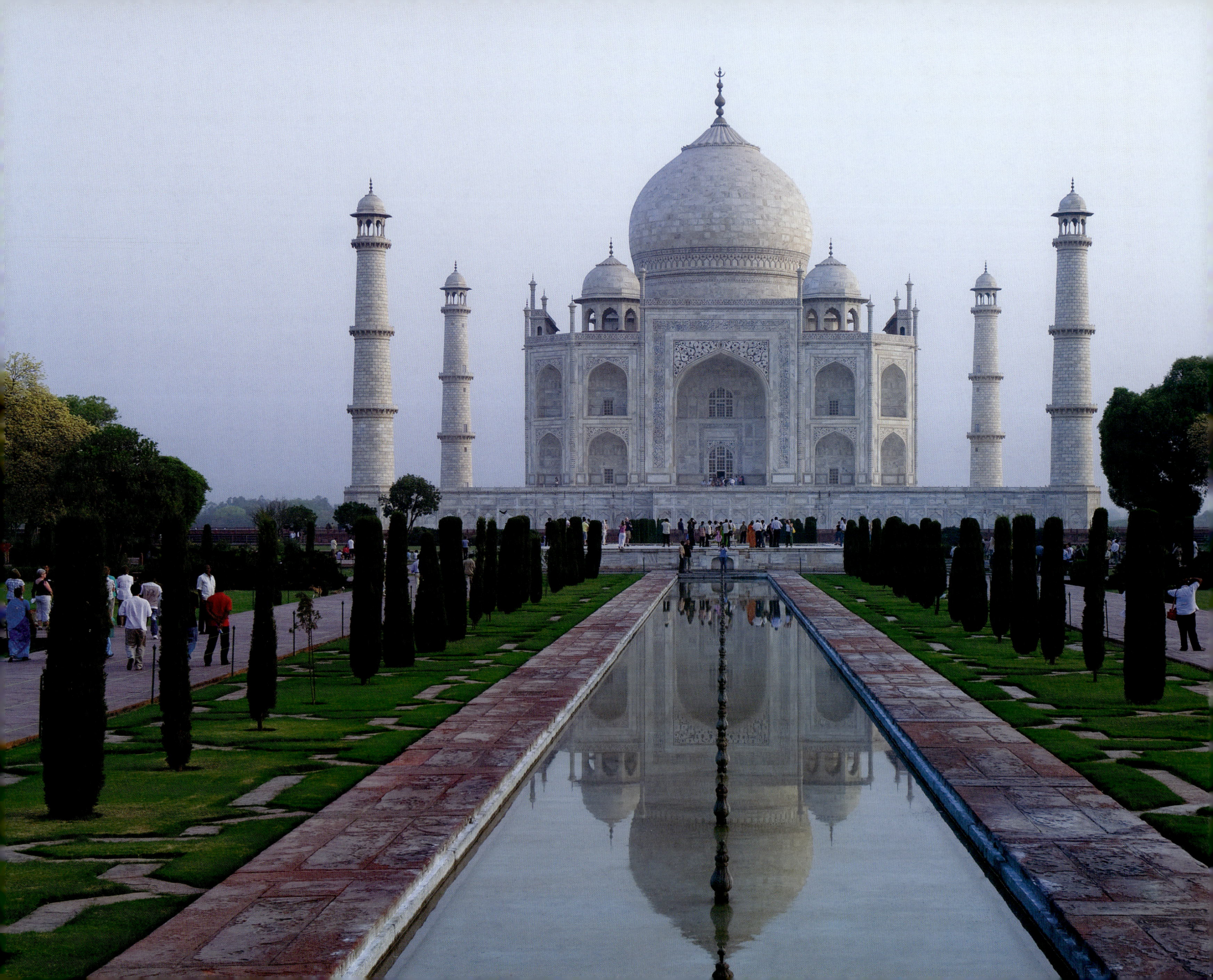

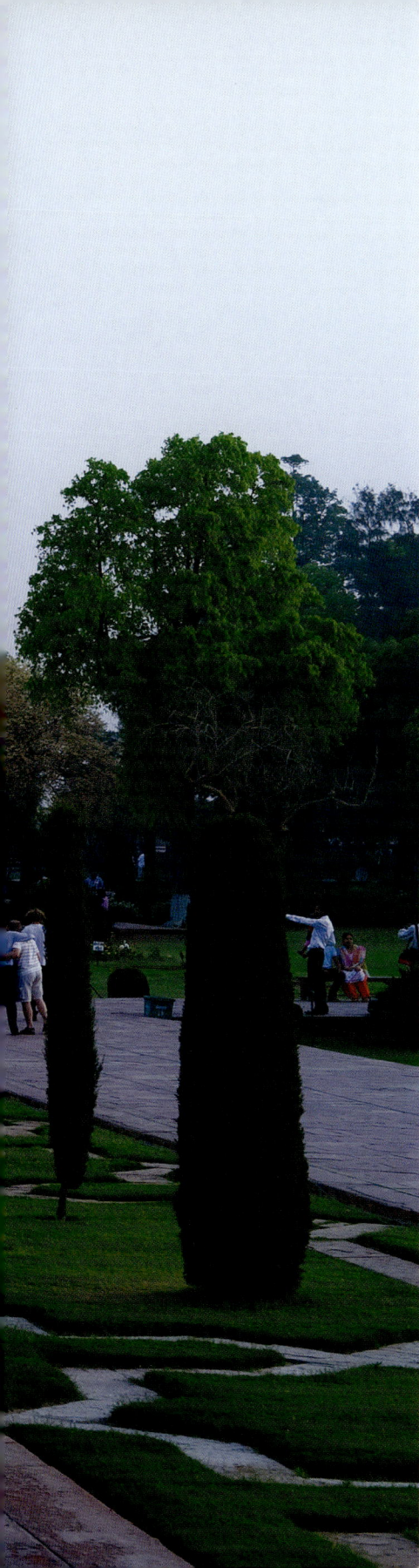

The Splendour of the Past

Monumental buildings remind us to this day of India's glorious past. Their grandeur and sheer size hint at the colourful and changeable history of the subcontinent.

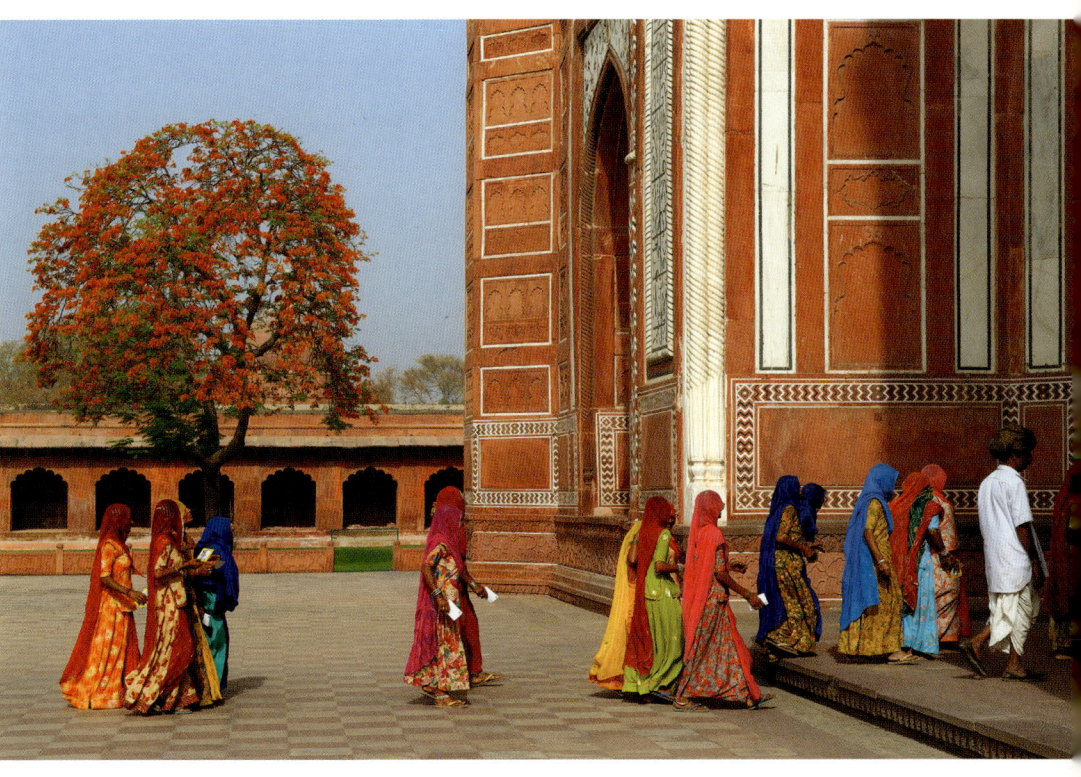

A Love Poem Made of Stone

At the crack of dawn a crowd has already started to form in front of the gates of the Taj Mahal in Agra. It is a mixed bunch, people of all ages and nationalities, united in their desire to see India's most famous building with their own eyes. While I queue for my ticket, persistent street hawkers try to sell me overpriced souvenirs. Then hordes of self-appointed tour guides offer their services – I politely refuse. After a very formal but seemingly rather ineffectual security control, I am allowed to enter the area. It is a whole complex of buildings, far larger than I had expected. As soon as I step through the gates I am looking straight at the Taj Mahal. It is mirrored in a large pool. An amazing sight! Without question this is one of the most splendid buildings mankind has ever created. But such splendour came at a high price.

The Mughal emperor Shahabuddin Muhammad Shah Jahan had the Taj built as a mausoleum in remembrance of his favourite wife Arjumand Banu Begum. She carried the title of Mumtaz Mahal – the chosen one of the palace – and had died during the birth of her fourteenth child in 1631. The story goes that her husband's hair turned grey overnight. Work on the building, which was designed to reflect the beauty and grace of his beloved wife, began the same year. It was a gigantic undertaking: some 20,000 men are said to have taken 17 years to build it, with the help of a thousand elephants. The best craftsmen worked with the most precious materials. Artists came from all over to fashion ornaments out of marble and countless semi-precious stones. The cost of the magnificent building increased so astronomically that national bankruptcy threatened. Riots and uproar broke out, and soon Shah Jahan was ousted by

his son Aurangzeb. The overthrown ruler spent his final years in a prison overlooking the Taj Mahal – his love poem made of stone. Yet he still achieved his goal, for the architectural masterpiece immortalized his wife.

But not all buildings are as well preserved as the Taj. The temple complex of Khajuraho in the Indian state of Madhya Pradesh, for example, spent a long time in a Sleeping Beauty slumber. The site was important during the rule of the Chandela Dynasty, from the tenth to the twelfth century, but after their demise the temple complex was reclaimed by the jungle. It was not until 1838 that the British engineer T. S. Burt rediscovered the buildings. He recognized the uniqueness of the elaborate filigree images that decorate the buildings, and the temple complex was once again made accessible. Of the 85 temples that had probably once stood at Khajuraho, 25 could be conserved. Today the Khajuraho temple complex is one of the major cultural attractions of India and is a UNESCO World Heritage Site – as are the ruins of Hampi.

From 1343 to 1565 Hampi was the capital of the Vijayanagara Empire, considered to be the last great Hindu empire of southern India. It was an important trading post for spices, fabrics and precious gemstones and attracted rich merchants. It is thought that Vijayanagara even then had a population of about half a million people. In 1565 the empire was defeated by an alliance of Islamic rulers, the Deccan sultanates. Several ruins attest to the splendour of this ancient metropolis. In recent years, people started moving back into some of the old buildings, and by 2012 the village of Hampi Bazaar had around two thousand inhabitants – the story continues.

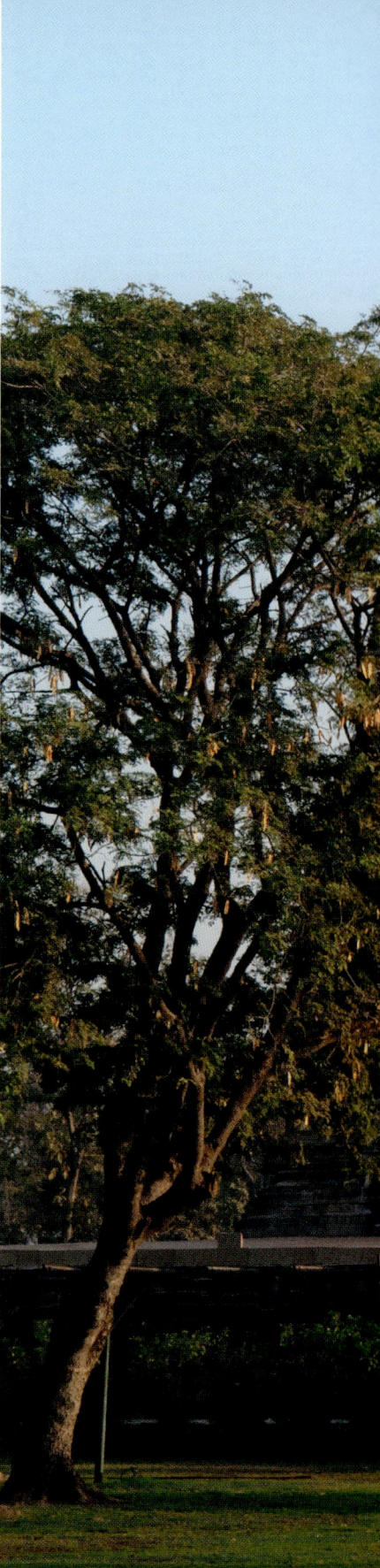

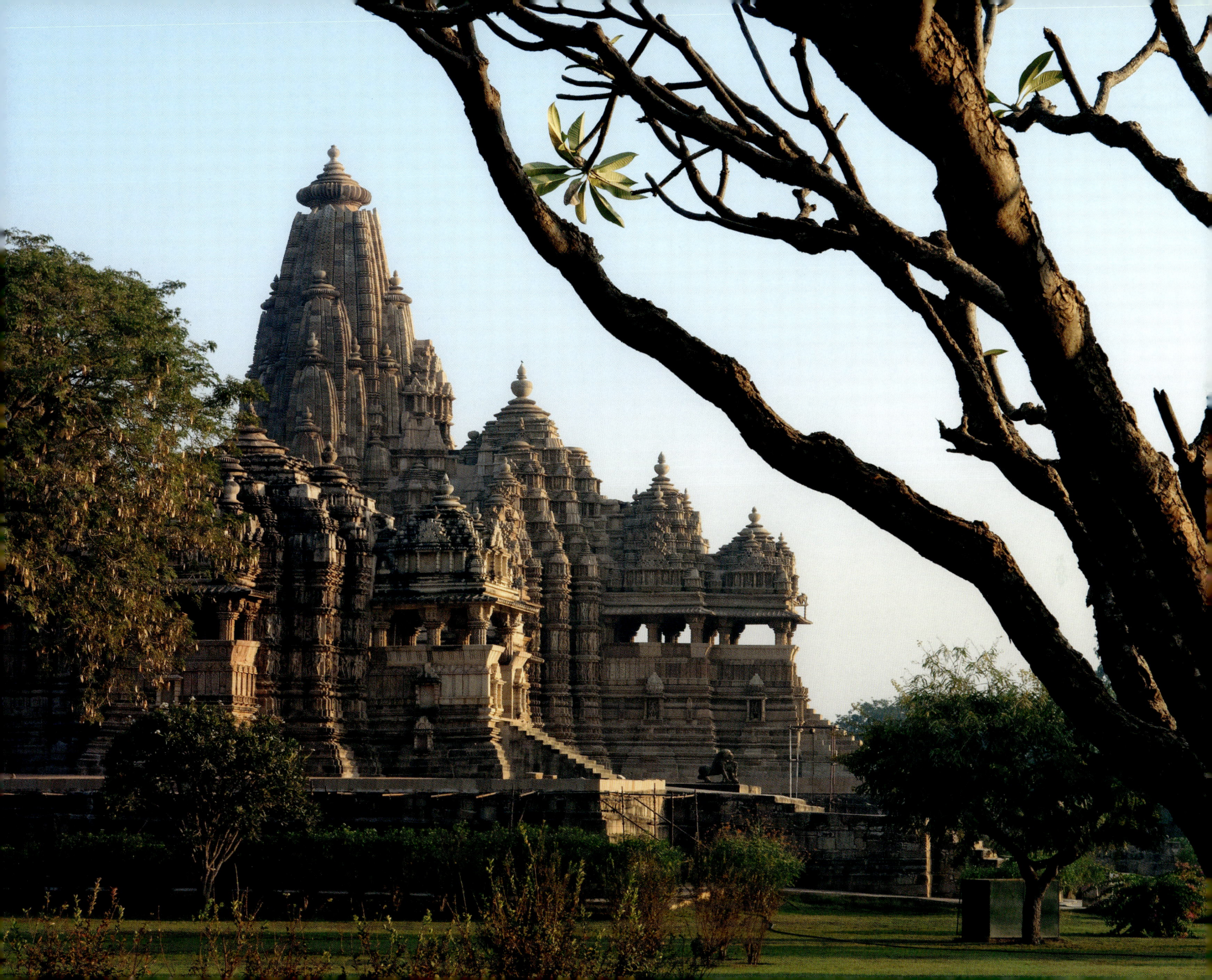

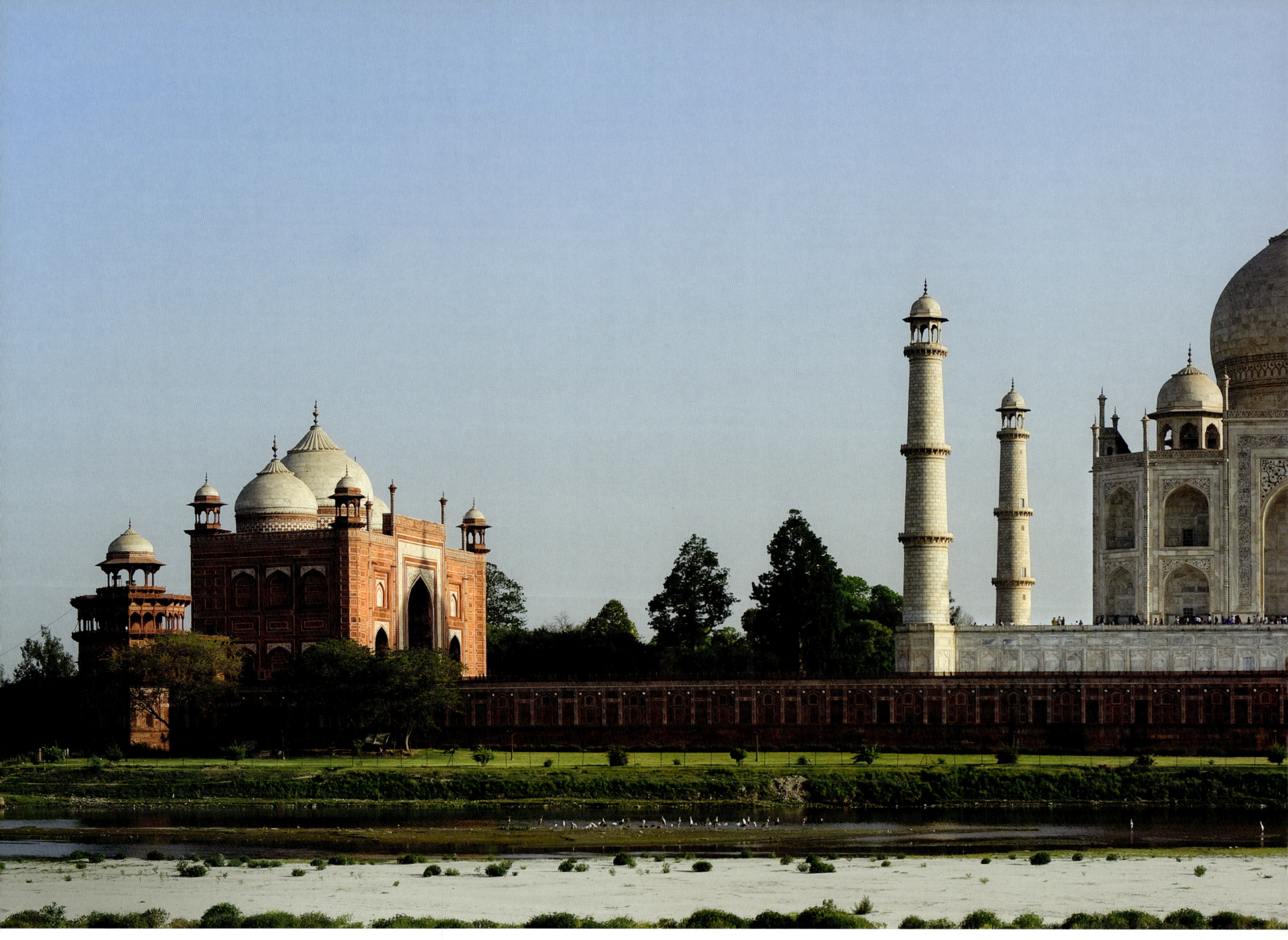

The Taj Mahal is built to a perfectly symmetrical plan. The view from the rear, over the Yamuna River, reveals two identical red buildings, one on each side. One is a mosque, but the other has no religious significance.

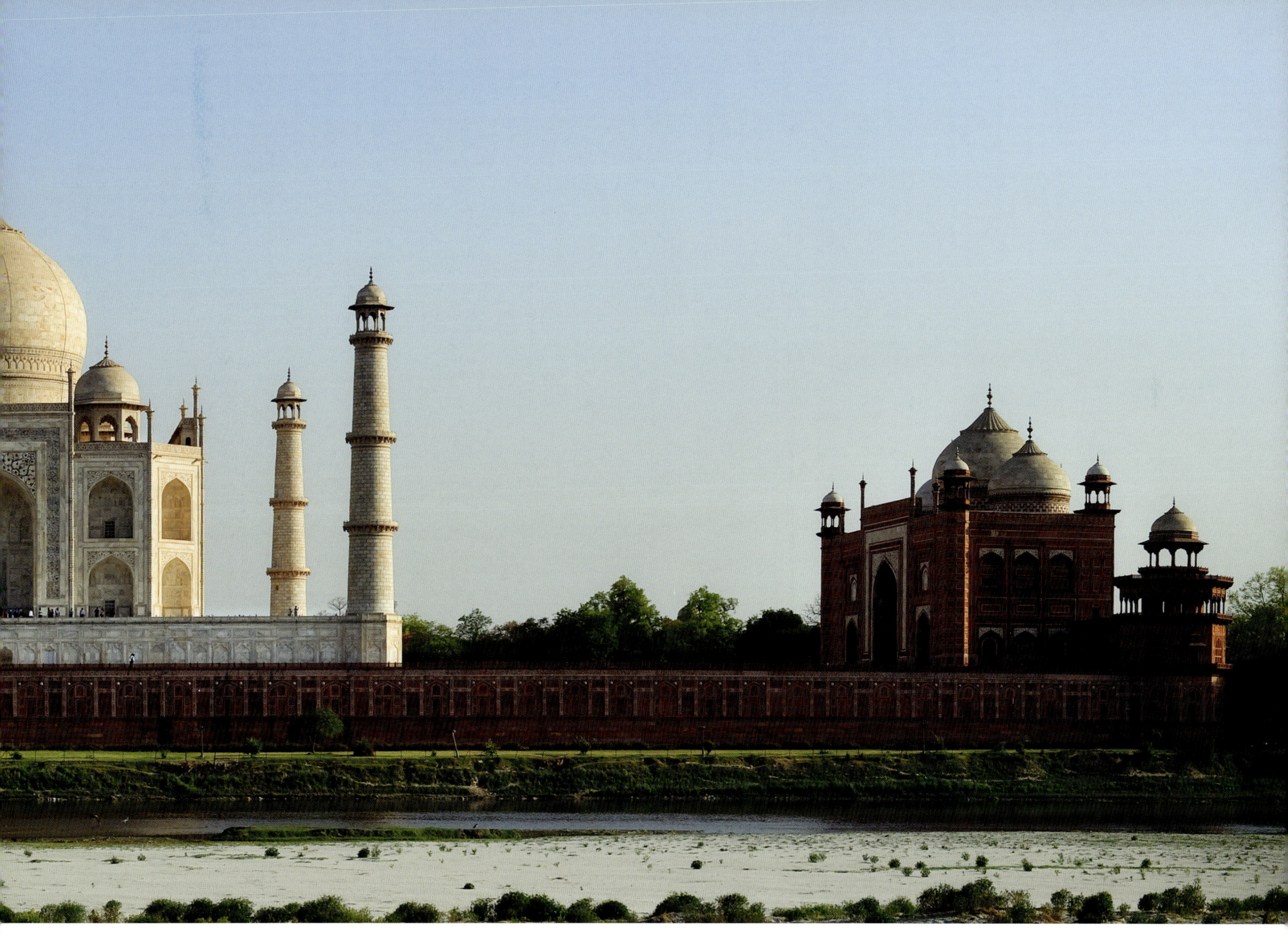

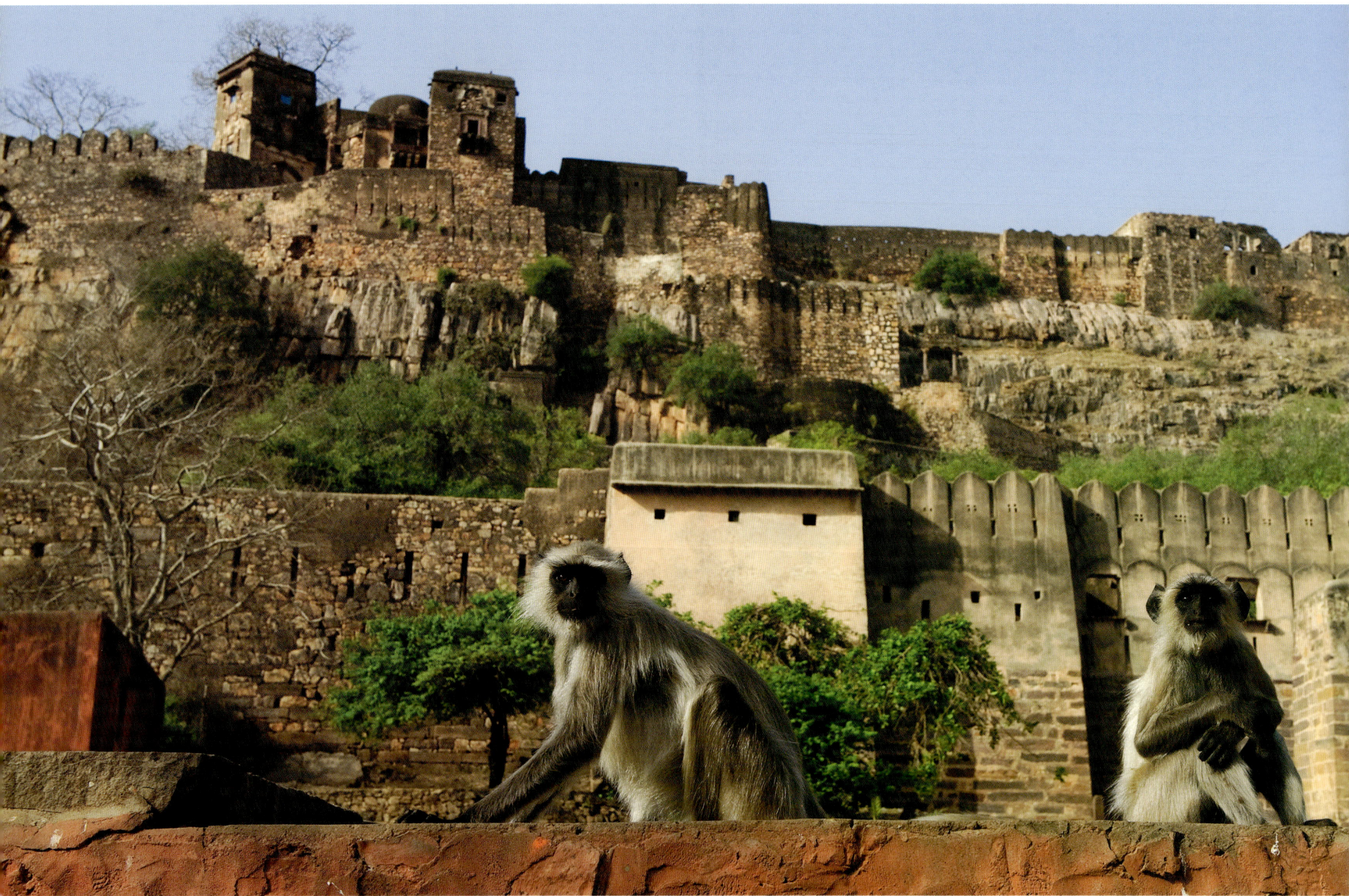

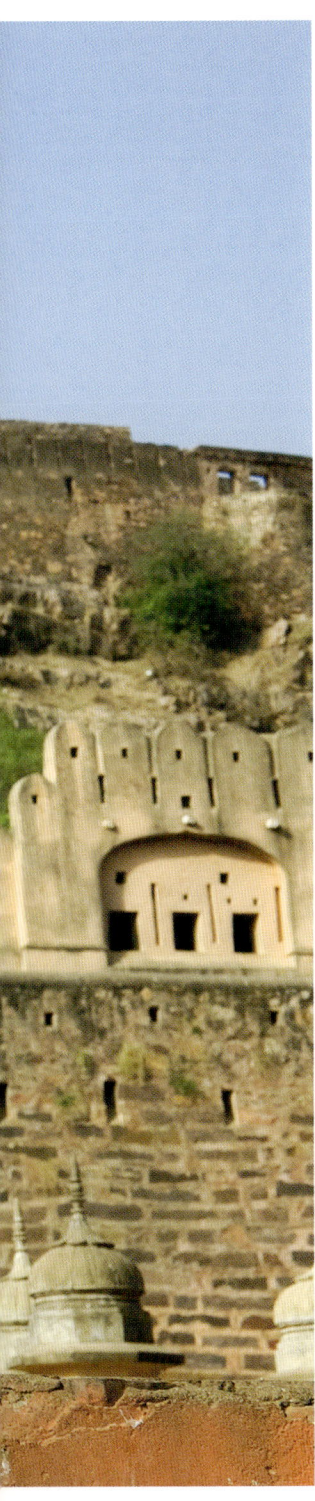

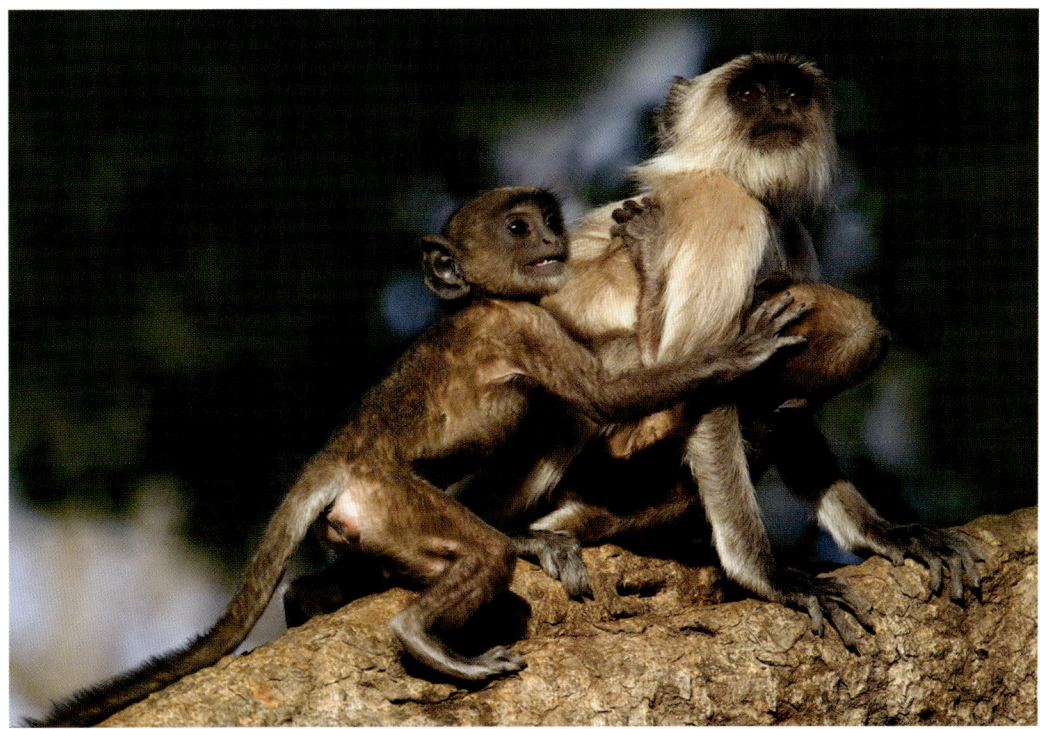

Rajasthan is home of the famous Ranthambore National Park, a prominent feature of which is an ancient fort of the same name. The ruins are inhabited by Hanuman langurs. They flourish here and are not shy, since pilgrims often feed them.

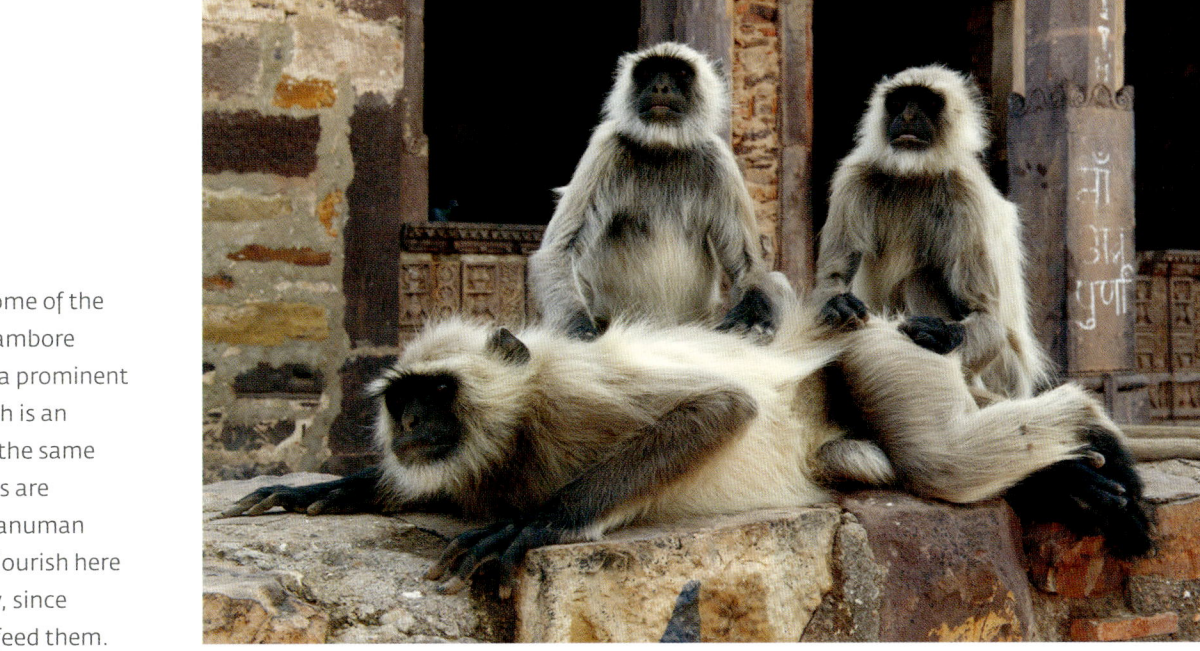

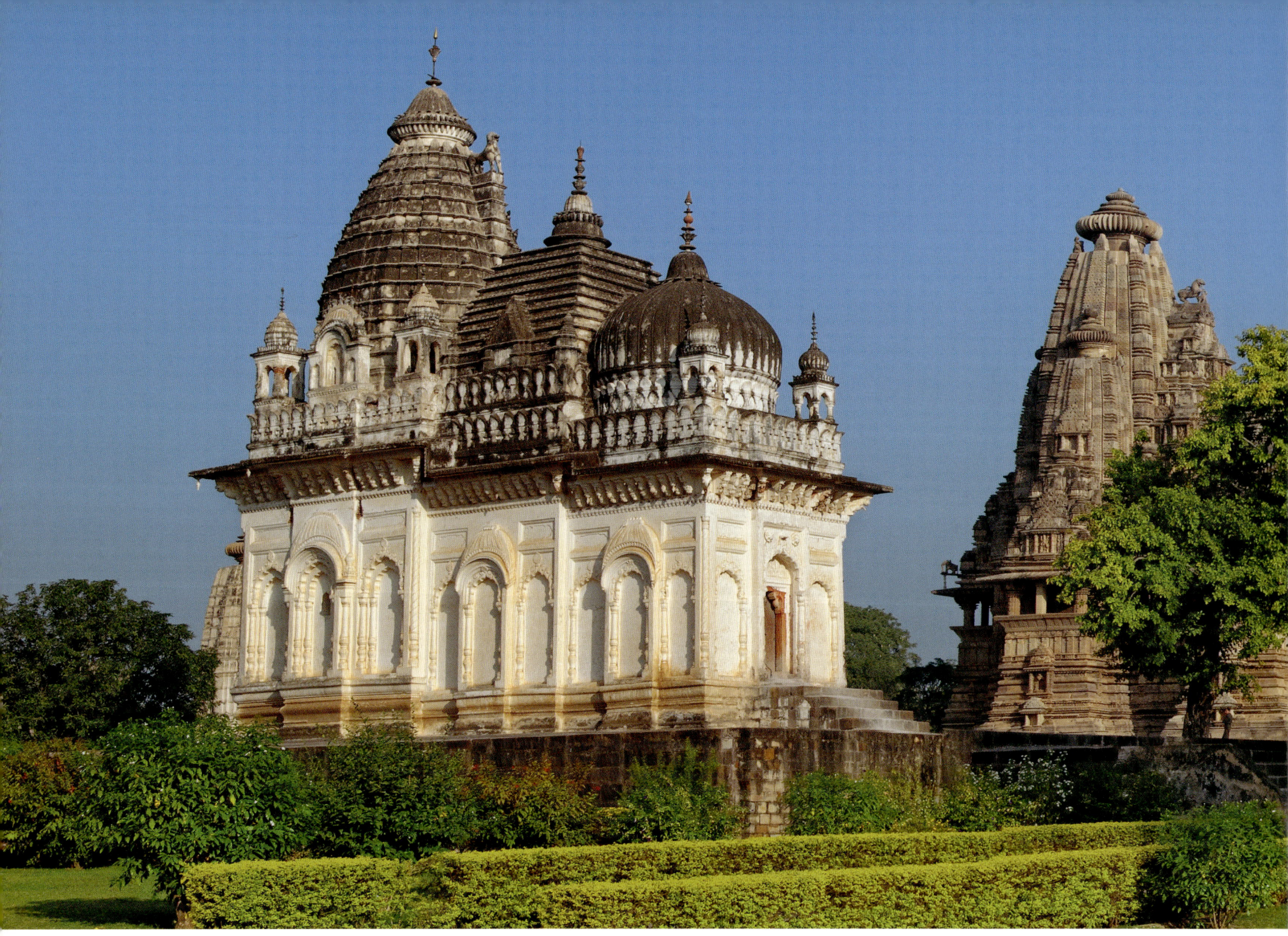

The temple complex of Khajuraho in the state of Madhya Pradesh is famous for its elaborate ornamentation.
For a long time it was completely forgotten, but it is now a UNESCO World Heritage Site.

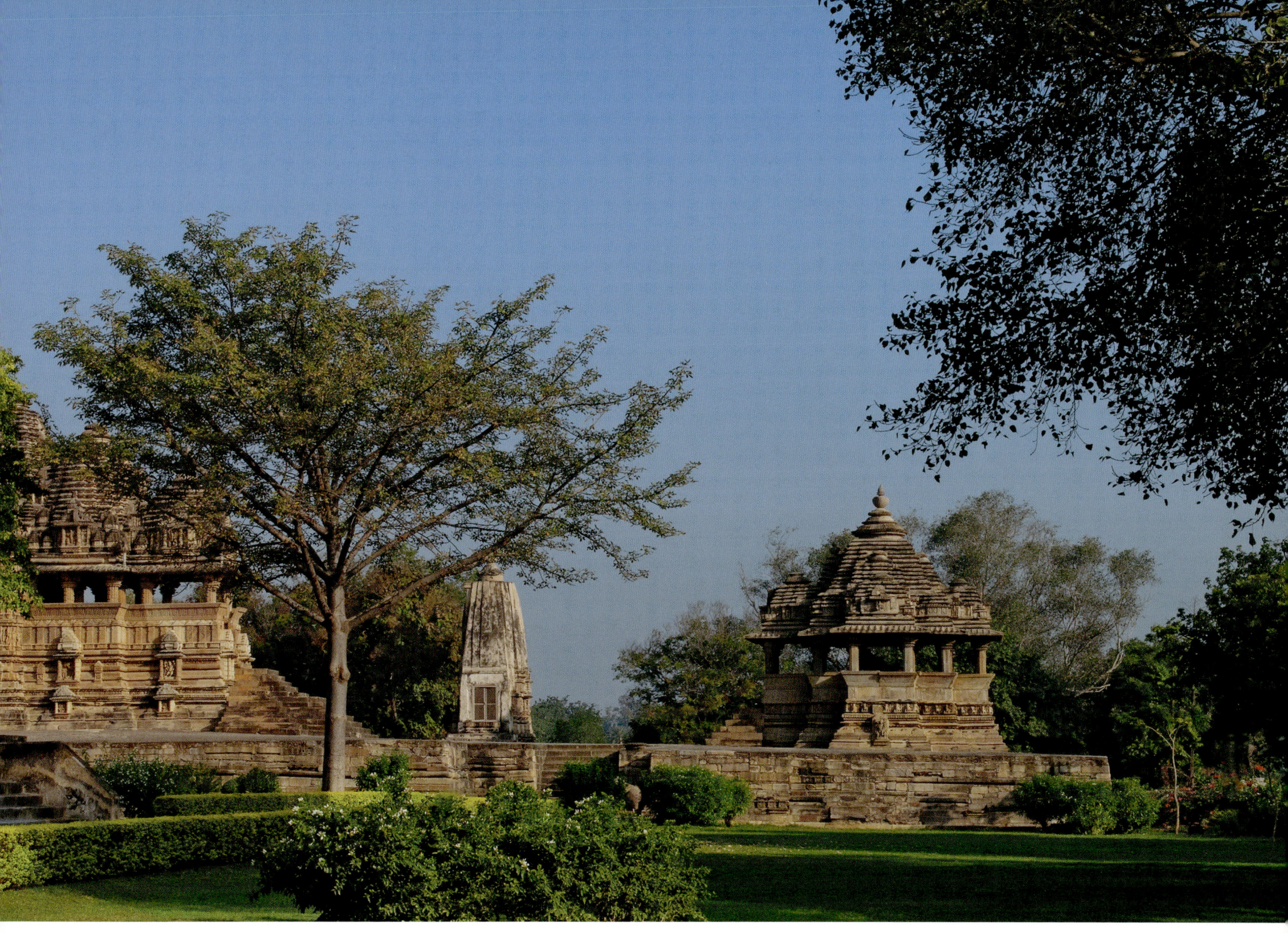

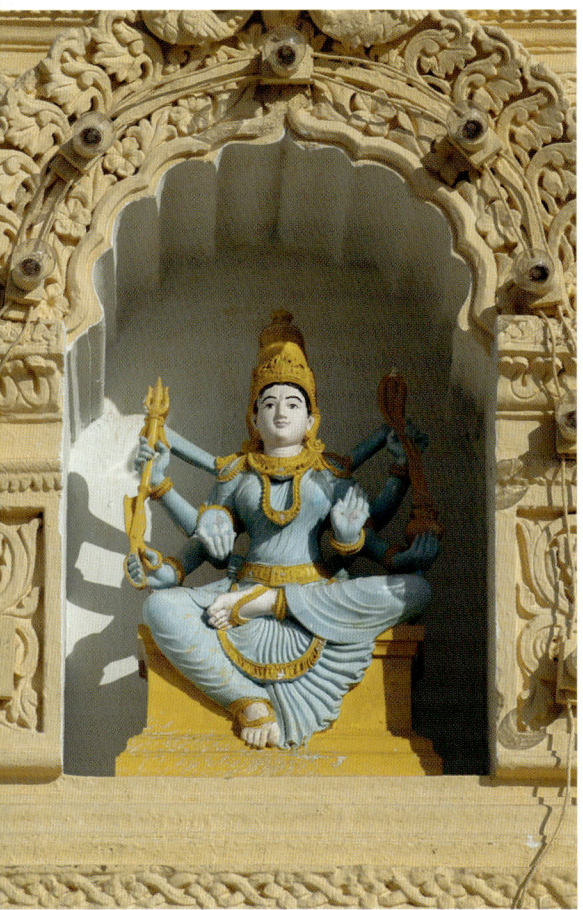
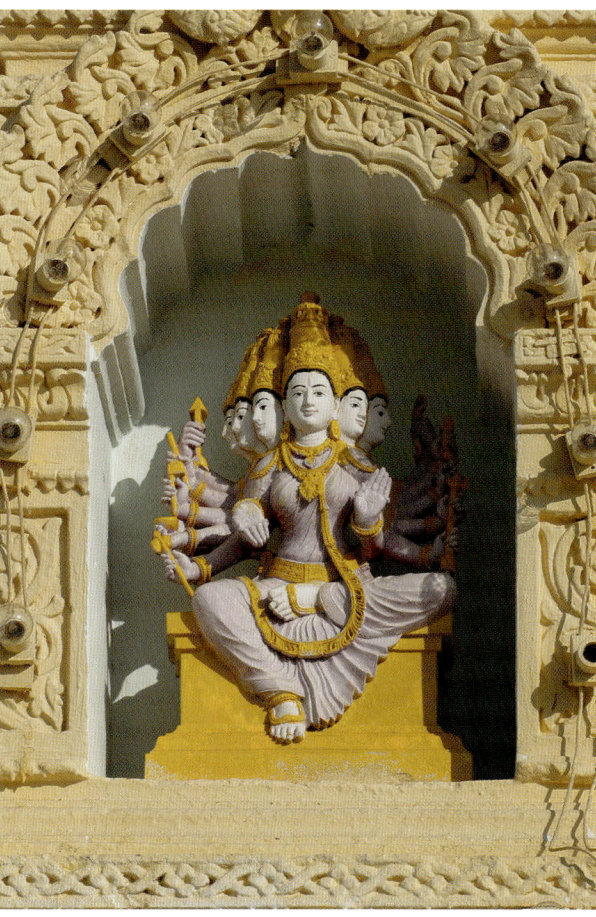
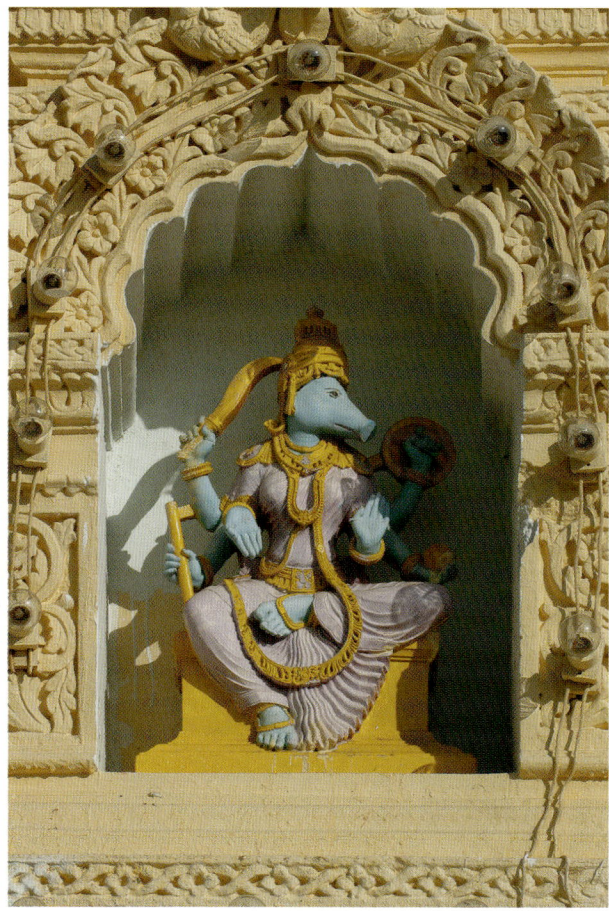

Mysore Palace was the official residence of the Maharajas of Mysore. It was completed only in 1912, after a previous building went up in flames. Decorated with numerous works of art and depictions of the gods, the complex is among the most famous palaces of India.

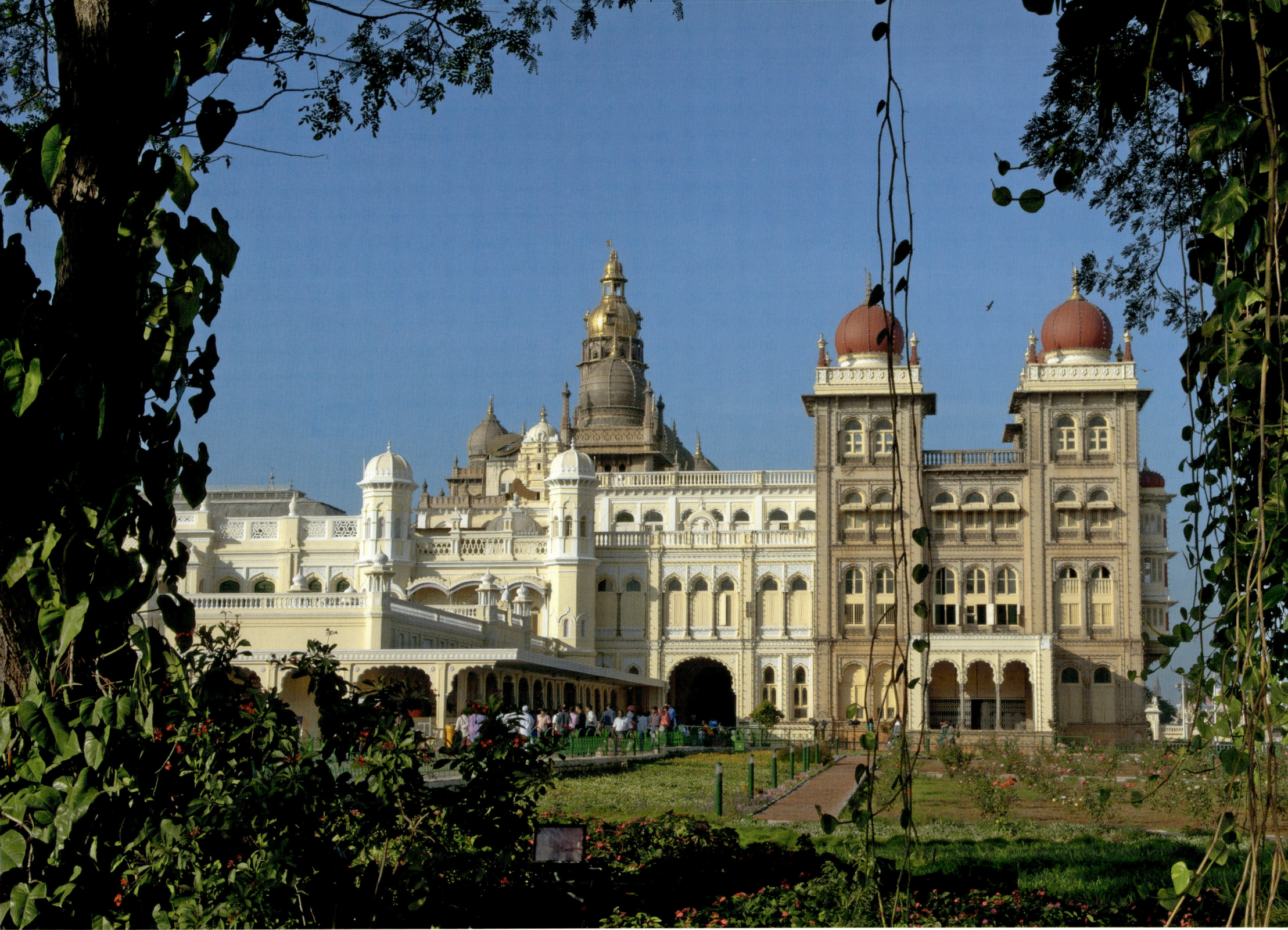

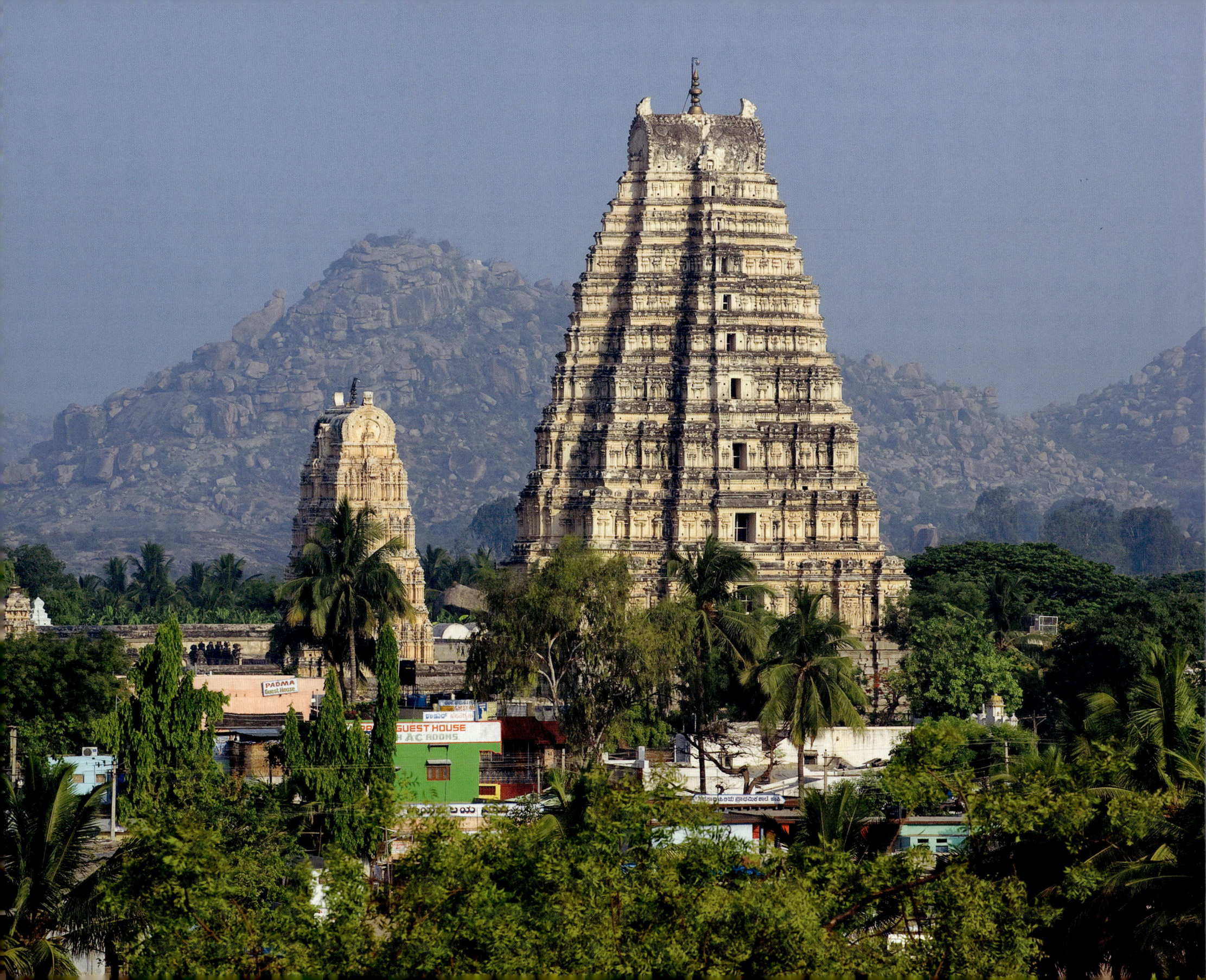

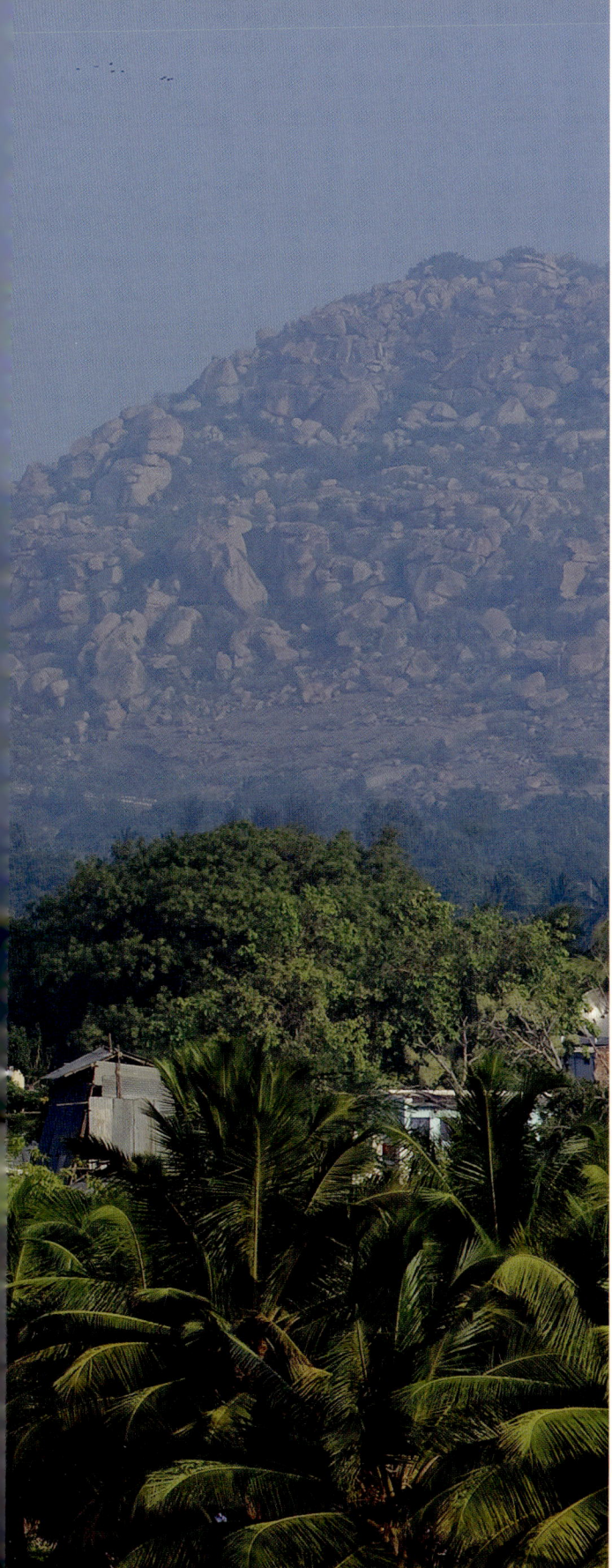

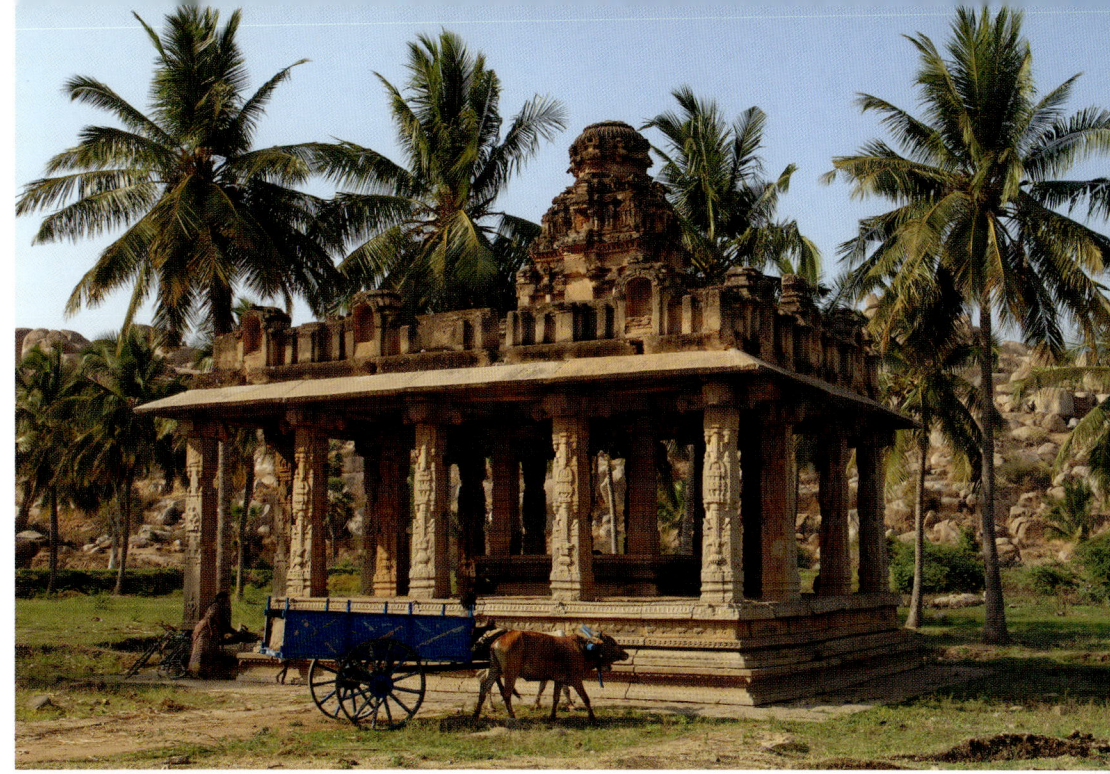

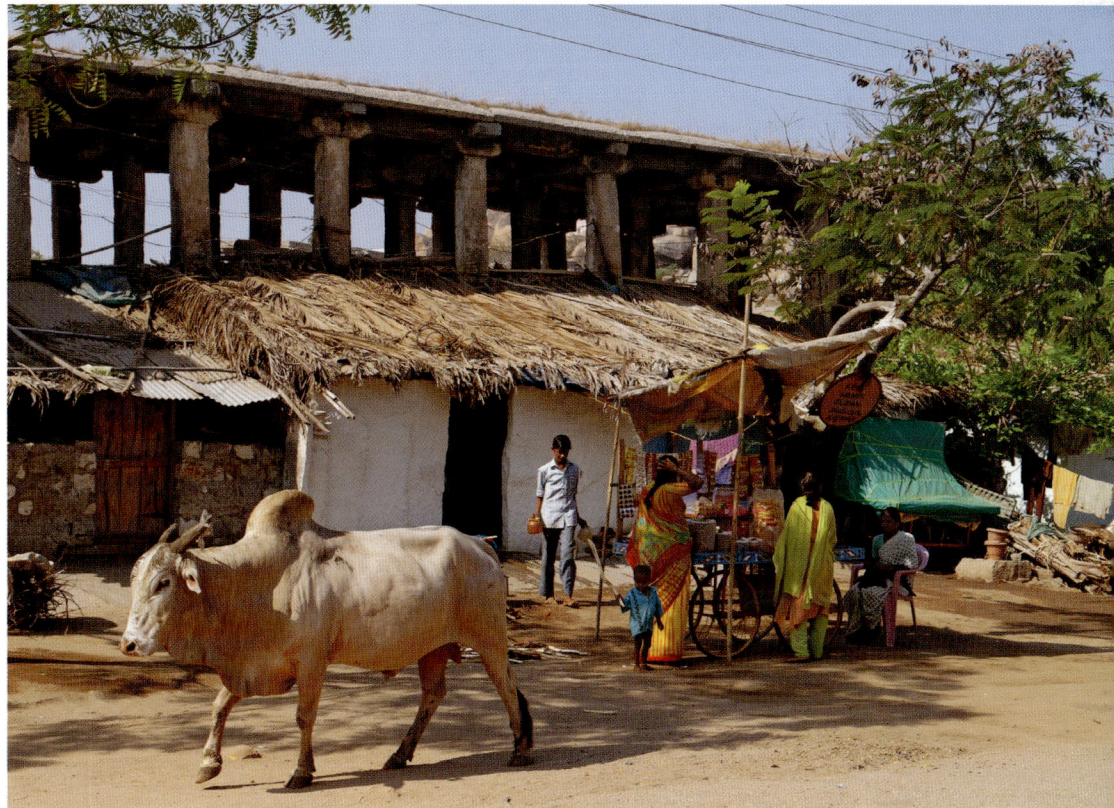

The ancient city of Hampi is dominated by a gopura, a gate tower that provides access to Virupaksha Temple (left). The whole area holds a number of old temples and other buildings. Several new houses have been integrated into some of the ruins.

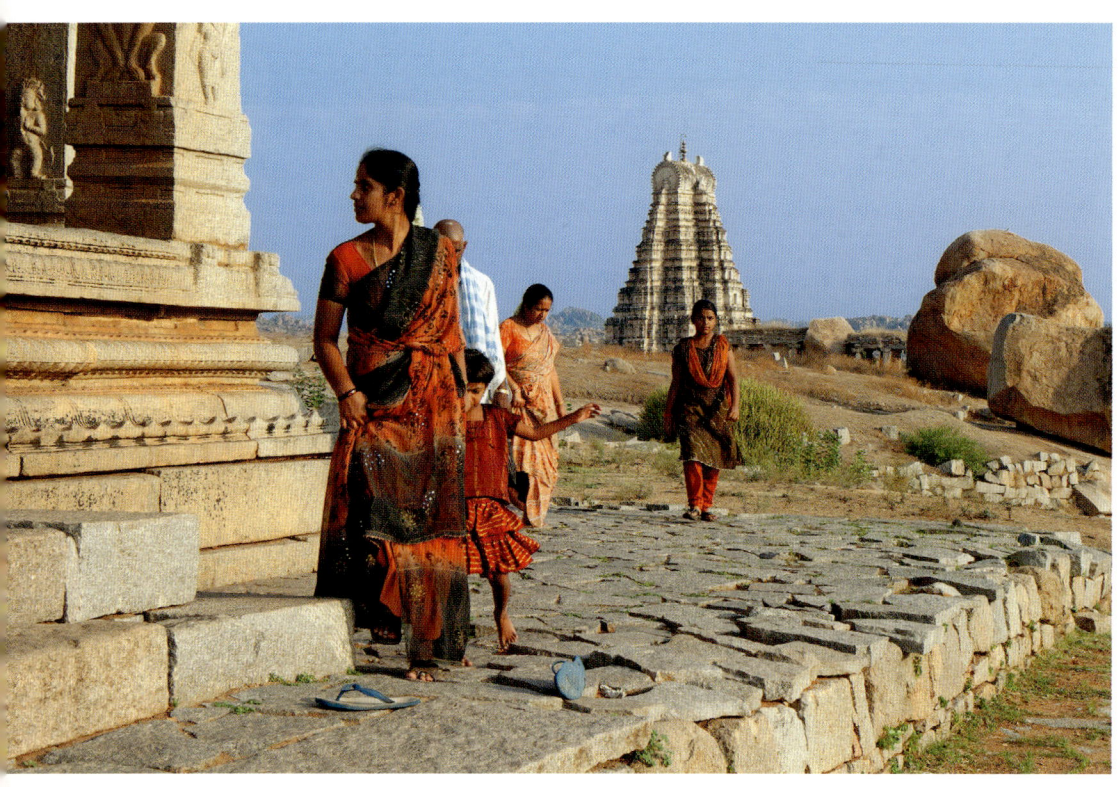
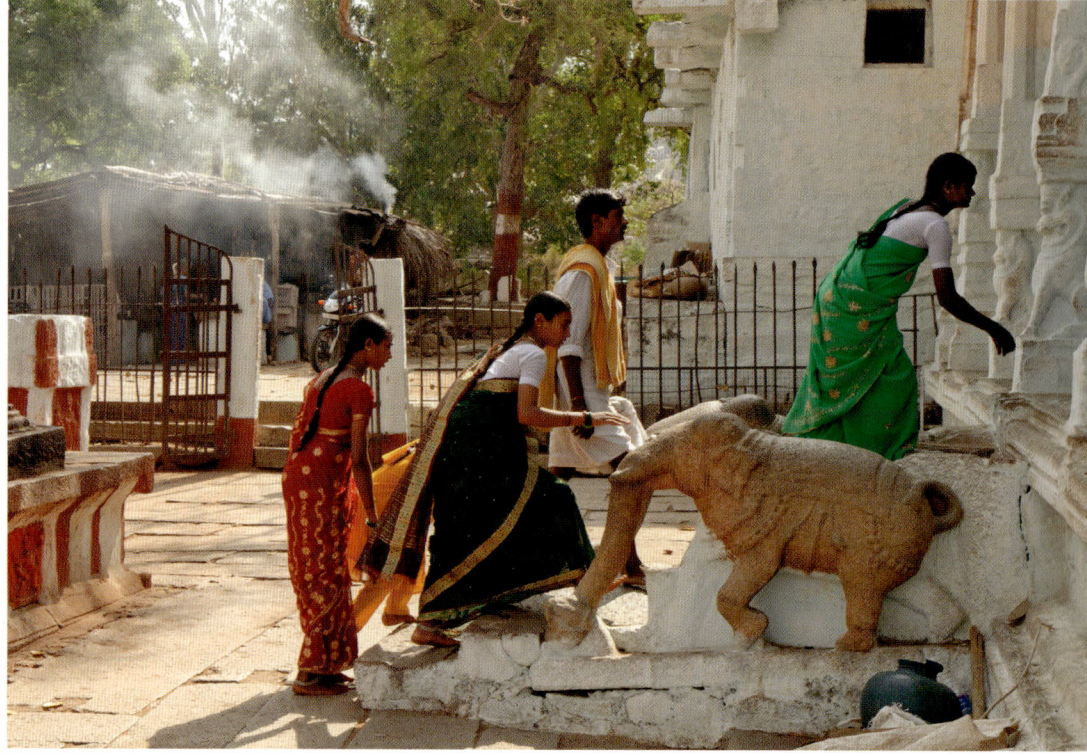

Hampi was once the capital of the thriving Vijayanagara Empire. Today it is located in the state of Karnataka. Pilgrims still like to visit the scattered temples of the complex.

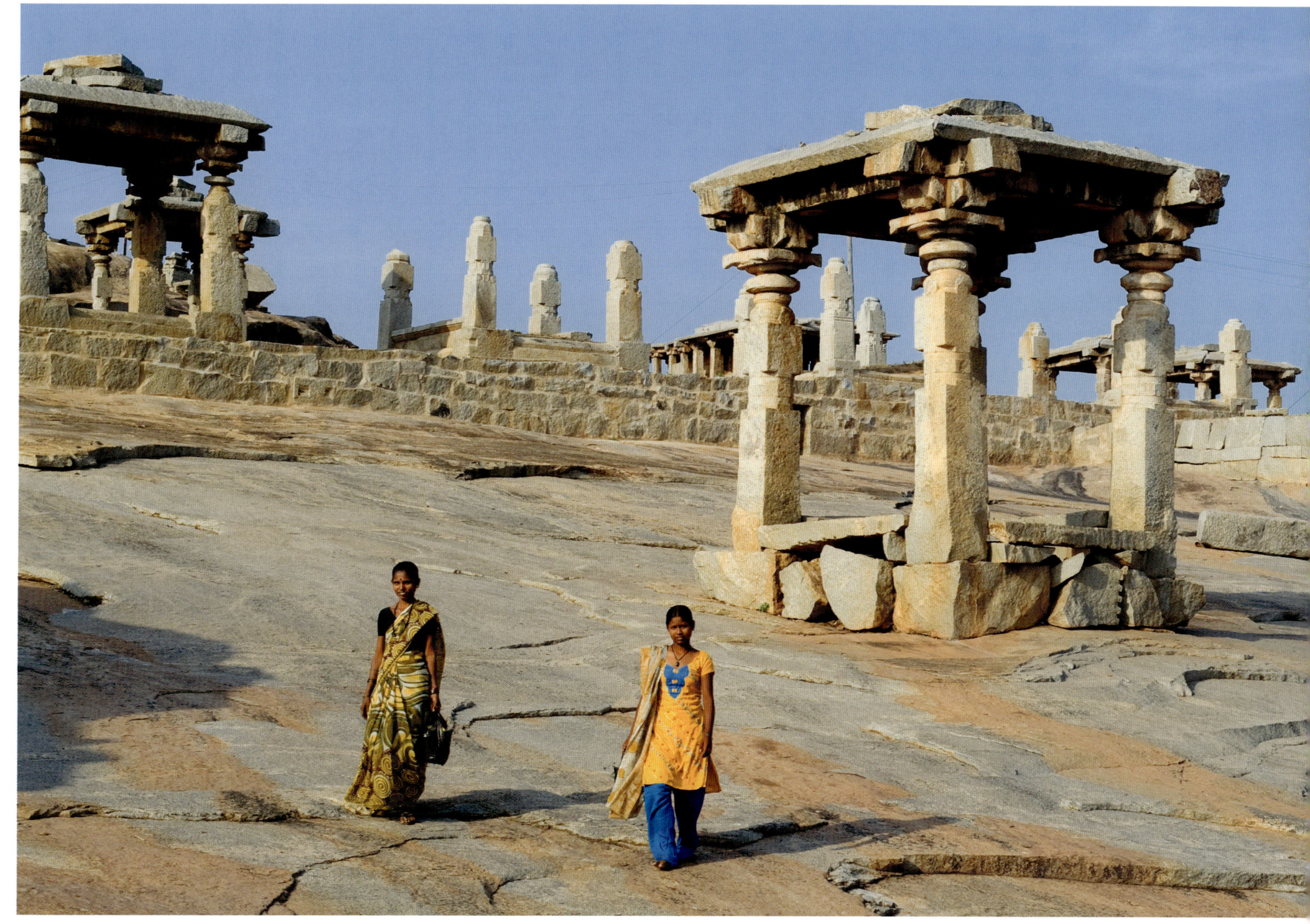

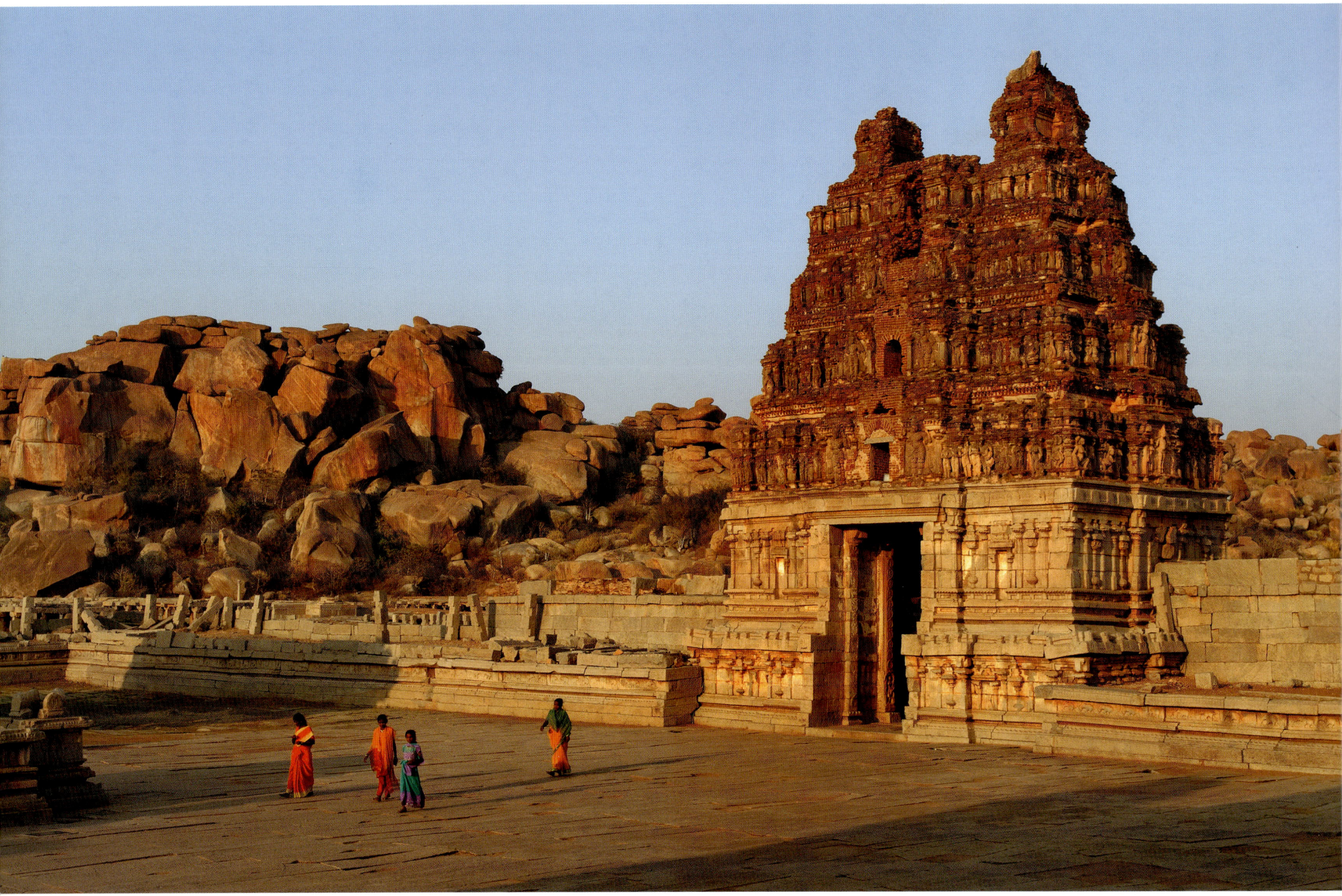

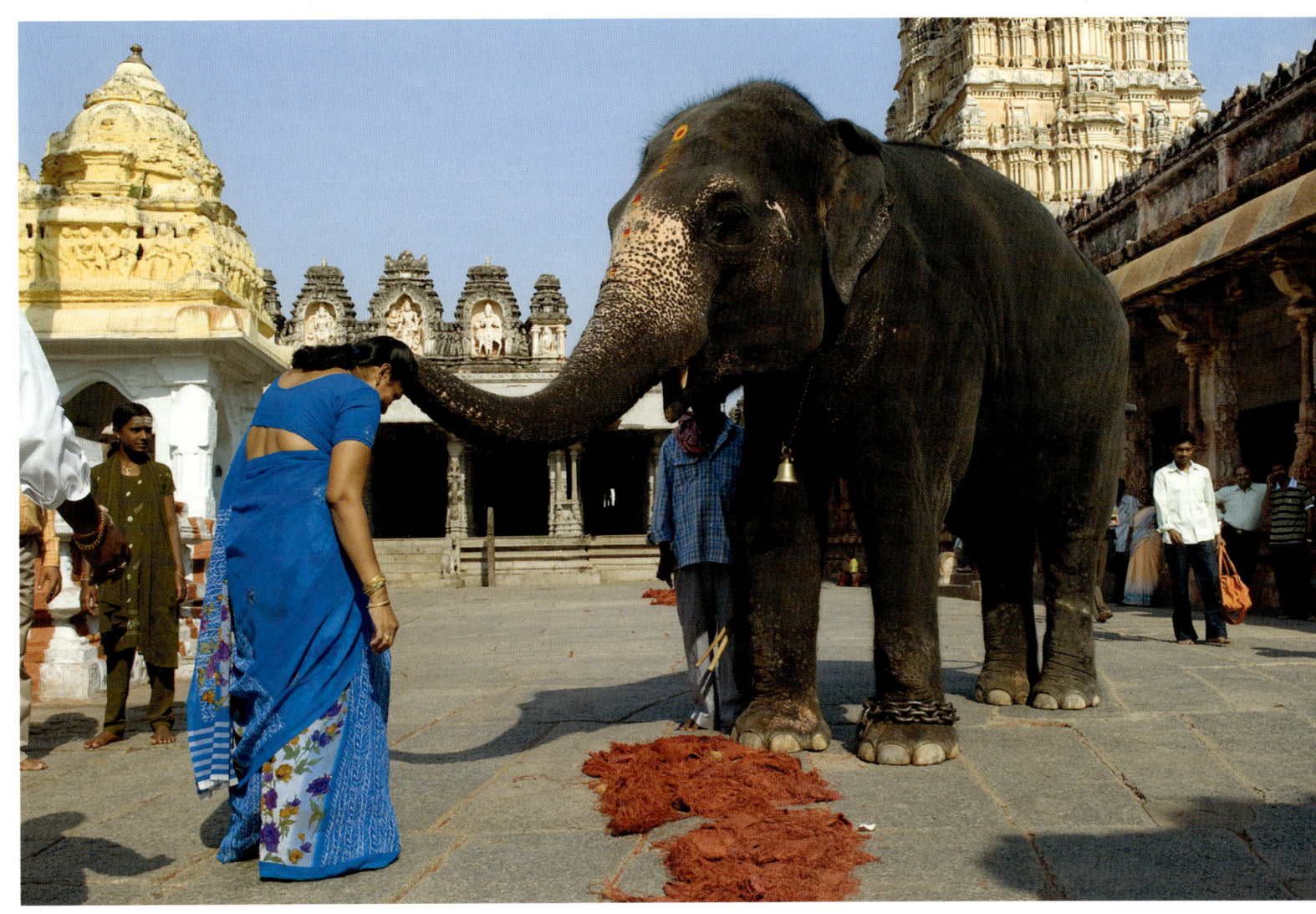

A number of well-preserved buildings in Hampi recall the splendour of the past.
This memory comes vividly alive in the court of Virupaksha Temple, where a
sacred elephant gives blessings to the faithful.

Acknowledgements

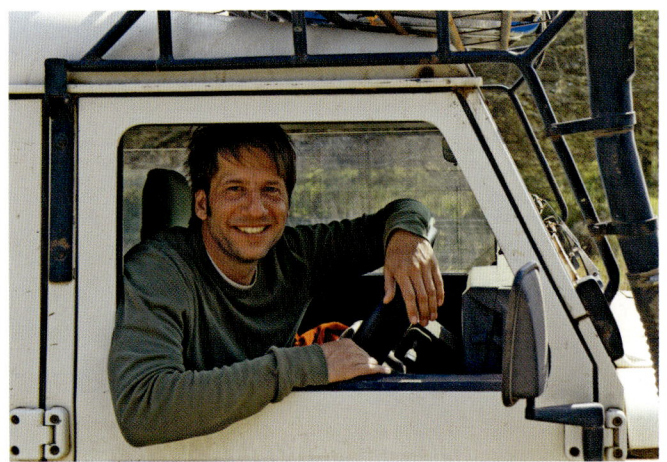

Over the years, many people have helped me to get close to tigers, sloth bears, elephants and many other fascinating animals. Among them were trackers, rangers, mahouts or villagers who lived close by. K. S. Abdul Samad and I. Ravindranath were particularly supportive. My colleagues Sandesh Kadur and Phillip Ross were extremely hospitable and showed me some of their favourite places in India, which in turn gave me the opportunity to take photographs in the wild. I am deeply grateful to all of them.

During my first trip to India in 1993 I met Kailash Sankhala. He was the first director of "Project Tiger" and has been very active in the conservation of these big cats. Later I worked for his son Pradeep Sankhala in the tiger reserves of Kanha and Bandhavgarh. I'd like to thank the staff of the local jungle lodges for both their hospitality and their support. Some time ago Amit Sankhala became managing director of Dynamic Tours. He is always an important and reliable partner for me in India.

I'd also like to thank Fritz Jantschke, since it was he who sparked my original interest in the country: he sent me to the subcontinent in 1993 to photograph for the magazine *Das Tier*. I'd been fascinated by the animals in Rudyard Kipling's *Jungle Book* since I was a child and now I was able to see them in the wild with my own eyes. I'm grateful to Gertrud Neumann-Denzau and her husband Helmut Denzau for the long-term exchange of information and ideas about Indian wildlife. During my early trips to India I met Valmik Thapar and Fateh Singh Rathore, whose work was always an inspiration. I'd like to thank the Tecklenborg family and their team for their support and the great production values of the original German edition and John Beaufoy who liked the book enough to produce this English edition – maybe because we both share an attachment to India.

My thanks also go to Yvonne Krüger who reviewed the original texts and provided moral support, when there just weren't enough hours in the day; to Hugh Brazier for polishing my English translation and to Rosemary Wilkinson for her project management. Finally, I extend my thanks to all those who contributed to the making of this book.

Axel Gomille

Axel Gomille
Further information : www.axelgomille.com